Trailblazing Women Empowered in the Industrial Revolution

Trailblazing Women Empowered in the Industrial Revolution

Paul Chrystal

PEN & SWORD
HISTORY

First published in Great Britain in 2026 by
Pen & Sword History
An imprint of Pen & Sword Books Limited
Yorkshire – Philadelphia

ISBN 978 1 03613 286 6

A CIP catalogue record for this book is available from the British Library.

Typeset by Mac Style
Printed in the UK by CPI Group (UK) Ltd, Croydon, CR0 4YY.

FSC www.fsc.org MIX Paper | Supporting responsible forestry FSC® C013604

The Publisher's authorised representative in the EU for product safety is Authorised Rep Compliance Ltd., Ground Floor, 71 Lower Baggot Street, Dublin D02 P593, Ireland.
www.arccompliance.com

For a complete list of Pen & Sword titles please contact:

PEN & SWORD BOOKS LIMITED
47 Church Street, Barnsley, South Yorkshire, S70 2AS, England
E-mail: enquiries@pen-and-sword.co.uk
Website: www.pen-and-sword.co.uk
or
PEN AND SWORD BOOKS
1950 Lawrence Road, Havertown, PA 19083, USA
E-mail: uspen-and-sword@casematepublishers.com
Website: www.penandswordbooks.com

This book is dedicated to Ada Anderson (1843–1913), fantastic pioneer of the popular sport of pedestrianism in the second half of the nineteenth century. There was nothing pedestrian about Ada.

Contents

Preface

We hear much about the contribution of men to the brave new world which began to emerge in the mid-eighteenth century with the start of the Industrial Revolution, and we can read the parallel story of the tens of thousands of women who clocked on at the mills, mines and factories to stoically power that revolution.

But much less is written of the women who were just as instrumental in these momentous developments and changes in our social and industrial fabric: these were the innovative and empowered women who pioneered developments in engineering and science, business, nursing, pharmacy and medicine, education and literature, music, art, politics and women's rights, labour law and the trade unions, philanthropy and entrepreneurship.

This book changes that great cover-up, that virtual erasing of many women from our history: *Trailblazing Women Empowered in the Industrial Revolution* is the rarely told story of 200 or so of these women, some familiar, but most, at best, languishing until now in the footnotes of British history. This is the story of how those 200 women, and others, helped power the Industrial Revolution and changed British society for ever.

It was never, however, the case that all men were intent on obstructing women's progress or achievements. Indeed, a number of enlightened men did acknowledge wholeheartedly the contribution women could and did make, for example, in business management, education, social reform, engineering, science, medicine, and in the arts – an important factor in women's fight for empowerment. As was the battle talented and innovative women of colour faced while overcoming male bias, sexism and racism; but some certainly did, despite the serried ranks of flat-earth men, and some flat-earth women.

Part One

The Social, Political and Economic Background

Introduction

The Changing Social, Political and Economic Background 1760–1900

When we think about the Industrial Revolution we think first of men, the men who harnessed the energy to power new machines and the men who built the factories in which to house them. But then there were the thousands of women who left the overcrowded homes and garrets that hitherto were their traditional places of work and moved *en masse* to the factories and mills which have provided, to some extent, their workspaces ever since. Sometimes, these women brought with them their children to complement the largely female workforce in specialised and often dangerous tasks – actually, a form of child slavery which was allowed to continue for many decades.

Received history then guides us to think of the wealthy capitalists who owned the factories, the land, mines and mills and set the pay rates and conditions of employment; crucially, the same men, or men in their employ, did all the hiring and firing. Too many of them were cruel and unforgiving taskmasters, although we can begin to see the seeds of enlightenment and philanthropy emerging in the mid-nineteenth century in pursuit of a happier, and therefore theoretically more productive workforce which showed some consideration for the plight of factory women and their appalling working conditions.

But then we think of the women again – not those toiling away on the factory floor, leaving home at the crack of dawn, at the beck and call of the knocker-upper, clip-clopping to their mills in their clogs, working twelve-hour shifts then clocking off to attend to their second jobs that day, this time unpaid, looking after the man of the house, the (often) many children and the house itself – a model of insanitation, crowded, devoid of privacy in a dingy street often swirling with sewage, complete with outside toilet (shared) and possibly next door to a bloody, stinking slaughter house, noxious tannery or dye factory with its airborne stench and toxic chemicals. And then there were the crowd diseases, the miasmas and diseases which proliferated: typhoid, typhus fever, plague, smallpox, cholera and the like which studiously kept the life expectancy shockingly low.

No, these stalwart women are not our concern here – they are covered in my *Factory Girls: The Working Lives of Women and Children* (2022). Here we are concerned with a different breed altogether who between them did much to ameliorate the working lives of their industrial sisters, to educate the workers' children, to provide entertainment outside the shouty public houses, to fight for the right of women to vote, to invent machinery and tools that made work more efficient and altogether easier, to provide culture for the few who could read and write… It is these trailblazing, resilient and powerful women who are the subject of this book. Indeed, the 140 years of the Industrial Revolution gave women all manner of empowerment in so many different fields of endeavour. This is their story.

The years of the Industrial Revolution which spanned the reigns of George III (1760–1820), George IV (1820–1830), William IV (1830–1837) and Victoria (1837–1901) embraced a period of huge transformation and development in many aspects of British life as the nation moved inexorably into what we might term the early modern world. Massive strides were made in science, medicine and technology, in working life and industry, in the way we governed ourselves, in the abolition of slavery, how we got about and how we moved our goods domestically and for export, how we educated ourselves and our children, the myriad ways in which we entertained ourselves and communicated with each other. Hugely important were the changes in legislation relating to child labour, working conditions, poor laws and workhouses, women's suffrage and prison reform. This period saw the emergence of philanthropists and attempts to mitigate sinister life-shortening occupational diseases – and 'social' diseases. And then there was progress in the arts, employment rights, security at home and abroad and in the law and in the enforcement of that law. We fought seemingly endless wars to expand and consolidate our growing empire all over the globe.

Some of these women will be familiar, others much less so. For the familiar, the book reappraises their achievements and exposes the anecdotal and any mythology that has grown up around them, re-establishing the empirical evidence which supports the historical record. For the unfamiliar, their infrequency in the history books makes them no less significant – they have just suffered more than their popular sisters from the often whimsical historiography as written by men. Extensive research introduces their contribution to their given fields of excellence and so offers a more rounded and honest history of women during the Industrial Revolution. A history which all too often still reflects a decidedly male gaze.

The prevailing, and changing, social, political and economic climates in which our women excelled are detailed, in order to illuminate more precisely the impact of their achievements in a socio-political and economic context and the contribution they made to ensuing years and to what we might call modern Britain.

We emphasise the social constraints under which educated, intelligent, ambitious and enterprising women had to live and work, beset with prejudice, sexism, male chauvinism and arrogance; where education of women was restricted and academic achievement for many years went without reward or recognition; where the law regarded them as second-class citizens with few rights in the home, in the family or at work; where the privilege of voting was denied to them until after an almighty struggle; where the professions and financial institutions were out of bounds and where glass ceilings were the norm, jealously defended in bastions of male dominance by often insecure men anxious to preserve their stranglehold on the levers of power and influence.

Change

'The only constant in life is change.'
Heraclitus, sixth century BCE

'Intelligence is the ability to adapt to change.'
Stephen Hawking (1942–2018)

If there is one word which characterises the Industrial Revolution then that word is 'change'. Social, economic, cultural and political change. The catalyst for this was technical innovation on a remarkable scale which introduced steam-powered machinery not only to the workplace but in transportation too with the spread of a sprawling railway system. And with all of this came revolutionary new working practices with a very different workforce and radically changed business models. To sustain this there was a simultaneous demographic shift which took workers away from their agriculture work in the countryside into towns and cities and then out of the home (their traditional workplace) into the many factories and mills which sprang up to accommodate the machinery, the raw materials and the finished products.

Factory owners immediately recognised women as a viable means of maintaining high enough productivity and increasing margins. Women were usually cheaper to employ than men and were generally less disputatious, less likely to join any workers' activist movements such as Chartism, which first emerged in 1836. They might bring their young children into work where they could be exploited even more than their mothers; the women and children possessed fingers which were usually more nimble than the stubby digits boasted by their male counterparts – a significant asset in an industry where the machines were prone to frequent breakdown and the eye of a needle was king. Women were often subject to verbal and sexual abuse, having to navigate their way around the discomforting and frightening offers of shifts for sex. The women were all too aware that the men could behave thus with virtual impunity and complaints might only led to dismissal, hungry children and the workhouse.

Most women in factories were younger than 30 years of age, and most of these were teenagers. 'A British survey undertaken in 1818 found that women comprised a little over half of the workers in cotton textiles.'[1] In Scottish factories, the figure was even higher. Furthermore, in Manchester, for example, 'the highest paid female factory worker made a quarter of what the highest paid male laborer earned'.[2] Many women were employed in mines both below and on the surface.

Their new type of working day required them to clock on at dawn and clock off some twelve hours later, returning home, sometimes with working children in tow, to prepare the evening meal, put those children to bed and maybe finish off the double shift of up to fifteen hours with a spot of housework or laundry. This was the double burden: six days every week, the one poorly paid, working dangerous machinery and plagued by physical, psychological and sexual abuse, the other unpaid and frequently accompanied by domestic abuse and inebriation. Serial childbirth added to the burden, as did the sinister chronic and creeping occupational diseases which came with the territory.

The demographics changed dramatically too. Our first census in 1801 revealed a lot about us: in the century from 1750 to 1851, Britain's population exploded from 6 million to 21 million. London's population expanded from 959,000 in 1801 to 3,254,000 in 1871; Manchester saw growth in 1801 from 75,000 to 351,000 in 1871. Other towns and cities witnessed similar exponential growth. Generally speaking, the 1851 census revealed that, for the first time, many more of us were living in towns and cities than in the countryside. By 1880, only 20 per cent of the British population could be described as living in rural districts.

Towns and cities generate more gregarious societies than rural areas so more young people are in contact with each other, leading to earlier marriages, which in turn caused the birth rate to rise. For example, according to Chris Shelley, 'In urban Lancashire in 1800, 40 per cent of 17–30-year-olds were married, compared to 19 per cent in rural Lancashire. In rural Britain, the average age of marriage was 27, in most industrial areas 24, and in mining areas about 20.'[3]

Their towns and cities were hopelessly polluted, insanitary and not always fit for human habitation; sewage, blood and guts from slaughtered beasts swilled around the lanes, back alleys and streets while toxic chemicals and infectious discharges leached into the water supply; graveyards were often overfull, body fluids leached into water courses and rotting corpses might be seen floating around after heavy rain.

Crime was endemic and there was little enforceable law and less order. Gambling, weekend bingeing and prostitution frittered away the Friday afternoon pay while the pawn shops flourished and the children and women of the house got hungrier still. Overcrowding was rife with families subsisting in damp, vermin-infested and foetid cellars or in rooms cohabited with other families. Outside toilets, middens and wells were shared between any number of families; rubbish was just dumped anywhere while disease proliferated and rodents enjoyed a field day every day. In 1837, 1839, and 1847, there were typhus epidemics; in 1831 and 1849, cholera struck.[4]

Worst-case scenario all this may be, but plenty of towns and cities bear witness to these lamentable living conditions: Manchester, London, York, Haworth, Hull, Leeds to name but a few: 'In Liverpool in the 1840s, 40,000 people were living in cellars, with an average of six people per cellar.'[5]

York also had more than its fair share of indigence, squalor and deprivation. Around the famous Shambles, Betty Petre kept her cattle in the court before slaughter in the Shambles; Mr Huber collected sheep's guts and washed them in a drain before exporting them to Germany to make fiddle strings. Other residents included a chimney sweep and a prostitute, referred to locally as 'an old knock'. It was this bottomless pit of penury which inspired Seebohm Rowntree to engage in in his hugely influential research project on poverty at the turn of the century.

Social reformer Edwin Chadwick helped on putting together the 1834 Poor Law, the thrust of which was predicated on the principle that making the provision of poor relief so unpleasant would deter all but the most desperate. In 1842 he self-published *The Sanitary Condition of the Labouring Population of Great Britain*, paying the costs from his own pocket as the Poor Law Commission wanted nothing to do with it.

Slowly but surely authorities were acknowledging the inextricable link between poverty and poor or non-existent sanitation, not least between chronic filthiness and cholera. It was only in 1848, after some pressure from the Health of Towns Association, and under the Public Health Act, that anything was done to mitigate this unremitting squalor, and that was all a bit half- hearted. The Central Board of Health was established but was restricted by limited powers and funding, leaving the implementation of the Act largely to all too often inert local councils who, in theory at least, were empowered to improve drainage, provide clean water, and remove refuse from streets and properties.

That Sunderland was where the 1831 cholera broke out first with 215 reported deaths accounts for the town's alacrity in its compliance of the Public Health Act and its urgent need for an inspector to be sent forthwith. Accordingly, Robert Rawlinson made the journey to Sunderland in February 1850. Rawlinson's stinging report was published in 1851 and copies were distributed throughout the kingdom, the initial impact being that the genie was now well and truly out the bottle and officials and the middle and upper classes could no longer hide behind the excuse of sloth, incompetence and ignorance regarding the plight of the living conditions of the working classes. Here is an extract from Rawlinson's vivid report:

> There is a most filthy place between two walls from 4 to 5 feet wide, behind John Street … called the Stinking Ditch, and it is literally so. This nuisance has not been abated, for no one will own it. A little from the Brandling inn is a cottage in a most unhealthy state; the water falling from a large building completely soaks through the walls, behind which is a stable in which sometimes pigs and rabbits are kept. The manure is allowed to be deposited for weeks together, until there are two or three cartloads. When I first visited this cottage I found it occupied by a very clean old woman and her husband; their bed-room was very offensive; they are both since dead of cholera…
>
> There are very few private drains communicating with the sewers. The ordure and offensive matters from the houses are generally brought out and thrown into the streets, and perhaps principally near the gratings or gully-holes, where it remains giving forth unhealthy effluvia, until removed by the scavenger or a shower of rain. Indeed it is difficult to say where the refuse water from some of even the better class of houses finds an outlet.
>
> The want of cleanly habits in the people adds to the accumulation of filth. Depositories for abominations of every kind are made in the most central places, and tend to the spread of all kinds of foul vapours and nuisances.

A sure way of getting an accurate picture of the precarious nature of work and the prevalence of vagrancy is to look at the admissions records of the local workhouse. And that is just what social reformer Henry Mayhew did while researching his groundbreaking *London Labour and the London Poor* (1851) when he analysed and reported 'an average drawn from the returns for seventeen years of the occupation of the persons admitted into the Asylums for the Houseless Poor'. He found that one in three inmates had previously been in factory employment; thirteen were charwomen and washerwomen, 38 were weavers; 273 were lace-makers; 274 furriers; 277 straw bonnet-makers; 277 trimming and button-makers; 350 were hatters; 2,047 were drapers and a huge 10,390 were milliners and dressmakers. These are, of course, traditional female occupations where redundancy booked the women an often one-way ticket into the local workhouse – their contribution to the family income thereby evaporated, tipping the family into a situation where it could no longer make ends meet. It is, of course, only part of the problem because traditional male occupations were also afflicted by the same misfortune with the same consequences. Here the highest numbers were racked up amongst inmates formerly working as artists (1,347), shipwrights (1,358), engravers (1,133), coach-makers (909) and leather dressers and curriers (802). Musicians came next with 730 and tanners with 643.

Women in the mines

Initially, the labour market created by the Industrial Revolution in the coal mines did not discriminate much between the work allocated to men and women, with women often taking on jobs which required hard physical labour. We learn that they toiled away as hurriers, hauling heavy coal carts through mine shafts, a task that also suited the many children whose diminutive size allowed them to squeeze into narrow, claustrophobic mine shafts from which their bigger brothers and fathers were physically excluded.[6] The alarming situation only ended after the passing of the Mines and Collieries Act 1842, one of the earliest attempts to regulate the workplace.

This excerpt from the *Great Britain Parliamentary Papers 1842*, Vol. XVI, pp. 24, 196 gives us a clear picture of the laborious role played by women in the pits:

> it is only in some of the colliery districts of Yorkshire and Lancashire that female Children of tender age and young and adult women are allowed to descend into the coal mines and regularly to perform the same kinds of underground work, and to work for the same number of hours, as boys and men …
>
> West Riding of Yorkshire: Southern Part – In many of the collieries in this district, as far as relates to the underground employment, there is no distinction of sex, but the labour is distributed indifferently among both sexes, except that it is comparatively rare for the women to hew or get the coals, although there are numerous instances in which they regularly perform even this work. In great numbers of the coal pits in this district the men work in a state of perfect nakedness, and are in this state assisted in their labour by females of all ages, from girls of six years old to women of twenty-one, these females being themselves quite naked down to the waist … Girls, [says the

> Sub-Commissioner J. C. Symons], – regularly perform all the various offices of trapping, hurrying filling, riddling, tipping, and occasionally getting, just as they are performed by boys.
>
> One of the most disgusting sights I have ever seen was that of young females, dressed like boys in trousers, crawling on all fours, with belts round their waists and chains passing between their legs ... When I arrived at the board or workings of the pit I found at one of the sideboards down a narrow passage a girl of fourteen years of age in boy's clothes, picking down the coal with the regular pick used by the men. She was half sitting half lying at her work, and said she found it tired her very much, and 'of course she didn't like it'. The place where she was at work was not 2 feet high. Further on were men lying on their sides and getting. No less than six girls out of eighteen men and children are employed in this pit ... In two other pits in the Huddersfield Union I have seen the same sight. In one near New Mills, the chain, passing high up between the legs of two of these girls, had worn large holes in their trousers; and any sight more disgustingly indecent or revolting can scarcely be imagined than these girls at work – no brothel can beat it.

The workhouse

The workhouse or poor house was a social institution of last resort with its origins in the acute labour shortages caused by the rapacious Black Death (1348–1349) in a bid to restrict the movement of labourers under the Statute of Cambridge 1388; the ultimate effect was that the state became responsible for supporting the poor, an onerous burden which was unsustainable by the eighteenth century. Mass unemployment at the end of the Napoleonic Wars in 1815, the advent of new, efficient machinery to oust agricultural and other workers, and a series of bad harvests all conspired to create need for legislation discouraging the provision of relief to anyone who refused to enter the terrifying portals of the local workhouse. 'Workhouse' is one of British history's great misnomers: work indeed was routinely done but it was of limited value: breaking stones, crushing bones for fertilisers, and oakum picking were as good as it got. As the inmates over time became older, feebler and increasingly incapacitated then so their productivity rapidly declined in line with the work actually carried out.

Our first reliable statistics for the numbers of workhouses comes from *The Abstract of Returns made by the Overseers of the Poor* – a government survey in 1776 which put the number of parish workhouses in England and Wales at more than 1,800 (about one parish in seven), accommodating more than 90,000 poor souls.[7] This explosion was triggered by the Workhouse Test Act 1723 which obliged anyone seeking poor relief to enter a workhouse and commit to a prescribed amount of work, usually for no pay (indoor relief); the Act helped reduce irresponsible claims on a parish's poor rate.[8] Nicholls records that:

> the Act repeated the clause of the Poor Relief Act 1691 that ordered that in every parish a book should be kept, registering the names of everyone receiving relief and the reasons why. No one else was permitted to receive relief (except in cases of disease, plague or smallpox) except by authority of a justice of the peace who lived in or near the parish or who was visiting during quarter sessions.

The Act claimed that 'many persons have applied to some justices of peace without the knowledge of any officers of the parish, and thereby, upon untrue suggestions, and sometimes upon false and frivolous pretences, have obtained relief, which hath greatly contributed to the increase of the parish rates'.

The growth in workhouse numbers was also bolstered by the Relief of the Poor Act 1782, as proposed by Thomas Gilbert whose Act was meant to allow parishes to share the cost of poor relief by collaborating to form unions, known as Gilbert Unions, to build and maintain ever larger workhouses to accommodate the elderly and infirm. Cost-cutting became such an obsession with some Poor Law authorities that they resorted to forcing husbands to sell their wives, to avoid them becoming a financial burden on the parish. In one case in 1814, the wife and child of Henry Cook, who were living in Effingham workhouse, were sold at Croydon market for one shilling (5p); the parish paid for the travelling expenses and a 'wedding dinner'.

By the 1830s most parishes had at least one workhouse, but many were ineptly managed. In his 1797 *The State of the Poor*, Sir Frederick Eden wrote:

> The workhouse is an inconvenient building, with small windows, low rooms and dark staircases. It is surrounded by a high wall, that gives it the appearance of a prison, and prevents free circulation of air. There are 8 or 10 beds in each room, chiefly of flocks, and consequently retentive of all scents and very productive of vermin. The passages are in great want of whitewashing. No

> regular account is kept of births and deaths, but when smallpox, measles or malignant fevers make their appearance in the house, the mortality is very great. Of 131 inmates in the house, 60 are children.

No wonder then that workhouses have earned a reputation for being dreadful, penal institutions which offered no hope of rehabilitation or reform; they were literally a one-way dead-end street and were to be feared and, where possible, avoided; restricted exercise apart, they offered little of value in terms of labour. But there were exceptions; the website https://brewminate.com/poverty-in-georgian-britain explains:

> But life in the workhouse varied enormously from parish to parish. Some workhouses were clean and comfortable havens for the poor. Many provided education, rudimentary health care and clean clothing. Others echoed to the sound of children playing, many of whom were placed in local businesses as apprentices, and most workhouses allowed visitors to come and go as they pleased. Other parishes – particularly in small rural communities – refused to build parish workhouses altogether owing to their substantial running costs. In many parishes 'outdoor' relief remained the chief means of assistance, administered to the poor on an individual basis.

However, the York workhouse may have been more typical where a number of the city's parishes set up a joint workhouse in 1768 to accommodate ninety paupers in a former cotton factory at 26 Marygate, working with textiles. York Poor Law Union was established in 1837 and took over Marygate: hygiene left much to be desired, according to this 1839 inspection, triggered after an outbreak of food poisoning from a vat of soup:

> In a room, however, in my opinion as a Chemist, ill adapted for the purpose of preserving meat, we found a Beast's Head at that time offensive to the smell … This room adjoins and opens into a short and very narrow yard considerably tainted with all the effluvia rising from some privies at one end … The head not having been previously cleaned, a quantity of unwholesome mucus was attached to it.

The Guardians nevertheless concluded that the soup had made inmates ill because 'the usual quantity of Potatos [sic] had been omitted making the soup too strong and rich'. The Poor Law Commissioners recommended a replacement for the Marygate workhouse, reporting 'a permanent reservoir of foul air'; and the privies were 'without exception in an offensive state'.[9]

The middle classes and the standard of living

An urban middle class emerged, accounting for around 25 per cent of the population by 1800; naturally, many of these vacated the noisome and polluted city centres for new, desirable properties in the suburbs which often came with a garden. Professionals such as engineers, scientists, merchants, lawyers and doctors and so on kept servants to look after their children, keep the home tidy, and cook meals. The middle classes used their disposable income to shop in the many new and elegant shops. The 1810s witnessed a revolution when new street lighting using coal gas made the streets safer to frequent at night, thus boosting the night-time economy by allowing the burgeoning clubs, restaurants, coffee and tea shops, theatres and other places of entertainment to flourish.

The year 1811 saw the formation of the National Society for Promoting the Education of the Poor – one of many organisations focused on the poor although, in this case, the main agenda was the dissemination of the Anglican faith. Nevertheless, this society and the burgeoning philanthropic organisations show us that there was real concern regarding the plight of the poor amongst the middle classes, intellectuals, artists and writers, that something had to be done about the cavalier use of labour and the venal pursuit of profits over people.

The standard of living did rise for most people during the Industrial Revolution, on average by around 30 per cent, but only from the 1830s did it percolate down to the lower classes. The growth and increasing ubiquity of the new media (newspapers, pamphlets, magazines and the like) helped to highlight and promulgate these issues, giving them greater exposure than ever before. Art and literature such as Joshua Reynolds's moving *The Age of Innocence* (1788) and Dickens's *Oliver Twist* (1837) helped to inculcate the notion that children should be protected and cherished, and that poor adults should be given a chance to better themselves and the lives of their children.

Philanthropy

Philanthropy aimed at the poor and destitute came in many forms and was conducted by such eminences as Jeremy Bentham, Edwin Chadwick, John Stuart Mill, the Cadburys, Booth, Nightingale, the Rowntrees, Lever, Mayhew, Dickens, Reckitt, Titus Salt and many others.

The garden village was a special type of philanthropy championed by a number of industrialists in the nineteenth and twentieth centuries, usually but not exclusively for the betterment of their employees. Titus Salt was one of these men who lifted hundreds of his workers from the industrial hell that was Bradford and relocated them in a purpose-built and sympathetically designed garden city. His good work is covered in my *Bridgerton Revisited: Women, Fashion and Society in the Regency Period (1795–1837).*

Henry Mayhew

Henry Mayhew (1812–1887), journalist, playwright, and advocate of and agitator for reform. He was a co-founder and joint editor of the satirical magazine *Punch* in 1841. He is best known as a social researcher, publishing an extensive series of newspaper articles in the *Morning Chronicle* that was later compiled into the three-volume *London Labour and the London Poor* (1851), a groundbreaking, revelatory and influential survey of the city's poor. A fourth, more statistical volume followed in 1861.

He was born in London, the thirteenth of seventeen children and educated at Westminster School before running away to sea. He then served with the East India Company. Avoidance of the payment of debts was another abiding preoccupation. The following passage gives an idea of what Mayhew typically found and what he subsequently wrote, in this case in Volume 1:

> The pavement and the road are crowded with purchasers and street-sellers. The housewife in her thick shawl, with the market-basket on her arm, walks slowly on, stopping now to look at the stall of caps, and now to cheapen a bunch of greens. Little boys, holding three or four onions in their hand, creep between the people, wriggling their way through every interstice, and asking for custom in whining tones, as if seeking charity. Then the tumult of the thousand different cries of the eager dealers, all shouting at the top of their

> voices, at one and the same time, is almost bewildering. 'So-old again,' roars one. 'Chestnuts all'ot, a penny a score,' bawls another. 'An 'aypenny a skin, blacking,' squeaks a boy. 'Buy, buy, buy, buy, buy – bu-u-uy!' cries the butcher. 'Half-quire of paper for a penny,' bellows the street stationer. 'An 'aypenny a lot ing-uns.' 'Twopence a pound grapes.' 'Three a penny Yarmouth bloaters.' 'Who'll buy a bonnet for fourpence?' 'Pick 'em out cheap here! three pair for a halfpenny, bootlaces.' 'Now's your time! beautiful whelks, a penny a lot.' 'Here's ha'p'orths,' shouts the perambulating confectioner. 'Come and look at 'em! here's toasters!' bellows one with a Yarmouth bloater stuck on a toasting fork. 'Penny a lot, fine russets,' calls the apple woman: and so the Babel goes on…

Arthur Munby

Arthur Joseph Munby (1828–1910) was a man of considerable talents; he can lay claim to diarist, poet, portrait photographer, barrister and social reformer. He is inextricably linked with Hannah Cullwick, about whom historian Leonore Davidoff writes that Arthur Munby's relationship with Hannah Cullwick was predicated on disrupting and redefining the established sexual and social norms of Victorian Britain. Their bizarre relationship also pushed the social boundaries surrounding race and empire through Cullwick's racial crossdressing. All of this allowed the couple to challenge and have control over the rigidity of Victorian society.

Joseph Munby was an extremely active professional, always at the centre of York life. Apart from establishing and running a prominent legal Gaol Sessions and Clerk to the Visiting Justices in 1845 he was a founder of the York School for the Blind in King's Court, Secretary to the York County Hospital, on the committee of York's Bluecoat School and York School of Art, a friend of York Ragged Schools and the Yorkshire Philosophical Society.

Arthur was the eldest of five children, born in York. He taught Latin at the Working Men's College for more than a decade and helped promote the Working Men's College Volunteer Corp (WMCVC), a response to the national call for Volunteer Rifle Corps (1859) to defend against a threat of war from Napoleon III.

According to the *Victorian Web* (https://victorianweb.org/authors/munby/bio.html) his dominant note may be said to have been what has been called 'the glorification of the working woman', with especial insistence on the dignity of manual labour.

So, Munby had a passionate, scholarly interest in working-class women, especially those who performed hard physical labour. We know that he often spent his time wandering the streets of London and other industrial cities where he stopped and questioned working women, enquiring about their lives and details of their work, while noting their clothes and dialects. These observations were recorded in his diaries, which feature sketches of working women. He collected hundreds of photographs, including women who laboured in collieries, kitchen maids, milkmaids, charwomen and acrobats. His diaries and images give us extensive historical information on the lives of working-class Victorian women, and our social history of the period is much enriched as a result. Much of his findings are published in his final book, *Faithful Servants: Being Epitaphs and Obituaries Recording Their Names and Services* (1891).

He noted in 1861 that on London Bridge:

> One meets them [working women] at every step: young women carrying large bundles of umbrella frames home to be covered; young women carrying cages full of hats, which yet want the silk and the binding; coster girls often dirty and sordid, going to fill their empty baskets, and above all female sackmakers.

Kathryn Hughes, in her *Guardian* review (18 January 2003) of Diane Atkinson's *Love and Dirt: The Marriage of Arthur Munby and Hannah Cullwick* (2003), reveals how

> Long before he spotted the 21-year-old scullery maid [Cullwick] in a London street, Munby had been trawling tens, hundreds even, of the lowest working-class women, including pit girls, mudlarks and, most exotic of all, trotter-scrapers. His motives were pure or purish. He didn't sleep with them, but instead, Mayhew-like, asked them questions about their living conditions, writing up their bemused answers at his Temple chambers where he was clinging on as a barrister. If he particularly liked a girl then he would pay for her to be photographed in her grime … Here, uniquely, is the engine room of high-Victorian life thrown open to the public gaze. And it is for this achievement, rather than as players in a peculiar sexual drama of their own

devising, that Hannah Cullwick and Arthur Munby are rightly remembered in this timely book.

As for Hannah Cullwick, Munby was taken by her size (5ft 7½ ins, 161 lbs) and strength, combined with the nobility of character he claimed to see in working women. Cullwick saw him as an idealised gentlemen who celebrated the intense labour she did as a maid of all work.

She was proud to call herself Munby's 'drudge and slave', and called him 'Massa', in a master/slave relationship. She typically wore a leather strap on her right wrist and a locking chain around her neck, to which Munby had a key. She wrote letters nearly every day to him, detailing her long hours of work. She would arrange to visit him 'in [her] dirt', parading the results of a full day's cleaning and other domestic work. She had a particular fetishist interest in Munby's boots, cleaning hundreds over the course of a year, sometimes by licking them. She once confided to Munby that she could tell where her 'Massa' had been by the taste of his boots.

Notes

1. Jeff Horn, *The Industrial Revolution*, Greenwood, 2007, p. 57.
2. Horn, 2007, p. 59.
3. Chris Shelley et al., *Industrialisation and Social Change in Britain*, 2016, p. 98.
4. See Paul Chrystal, *History of the World in 100 Pandemics, Plagues and Epidemics*, 2022
5. Benjamin Armstrong, *Britain 1783–1885*, London, 2020, p. 188.
6. Hurriers were workers who shifted the coal from the face to the horse-ways. Horse-ways were the main passages where horses could be used for hauling. Sometimes they used a pulley system to wind up the trams.
7. Peter Higginbotham, *The Workhouse Encyclopedia*, Stroud, 2012.
8. Cheryl Nixon, *The Orphan in Eighteenth-Century Law and Literature*, London, 2011, pp. 12–13, 57. Also known as the Workhouse Test Act 1722, or Knatchbull's Act: 'An Act for Amending the Laws relating to the Settlement, Employment, and Relief of the Poor'.
9. See Paul Chrystal, *Bridgerton Revisited*, Pen & Sword, 2026.

Chapter 1

Education For and By Women

The Industrial Revolution saw major changes in education generally, but education for women was significantly increased and extended. Nevertheless, the vast majority still saw very little or no educational provision and remained illiterate, with attendance at dame schools the best they could expect. Dame schools taught the very basics, and their role as schools was often blurred with that of childcare providers. The focus, if there was one, was on domestic skills in preparation for a life in domestic service or helping at home.

The National Society (Church of England) for Promoting Religious Education

The National Society was founded on 16 November 1811 and incorporated by Charter in 1817 as the National Society for Promoting the Education of the Poor in the Principles of the Established Church throughout England and Wales.

As such, it promoted the church schools and Christian education of the established church and competed with the nonconformist British and Foreign School Society. The National Society had the backing of the Anglican clergy, and Oxford and Cambridge universities.

Its aims and scope were:

> That the National Religion should be made the foundation of National Education, and should be the first and chief thing taught to the poor, according to the excellent Liturgy and Catechism provided by our Church.

It therefore became the main agency through which the Church operated before the 1870 Education Act – and it continued to be the Church's mouthpiece in education well into the twentieth century, acting on behalf of the Church in negotiations surrounding the 1870, 1902, 1918 and 1944 Education Acts.

The National Society was immersed in many aspects of education:

- The publication of books and the provision of equipment to the training of teachers.
- The founding of most Church of England/Church in Wales schools (National Schools). The aim was to found a church school in every parish in England and Wales by raising funds centrally and then offering grants to prospective founders on conditions that enabled the Society to influence and foster development of the scheme.
- The training of teachers to which end it ran training courses at its Central School in London and also founded St Mark's Chelsea, Whitelands College and Trinity College Carmarthen.
- Providing affordable books so that schools should have access to low-cost books. At first these were published by the Society for the Promotion of Christian Knowledge (SPCK). It also gave grants to schools towards the purchasing of books and equipment and set up a Depository in Westminster in the 1840s to supply these.

In 1840, 60 per cent of women remained illiterate, but by 1860 this had come down markedly to 40 per cent: a reflection of the limited liberating effect the Industrial Revolution had, releasing working-class women from the home to the factory bench or to the homes of the middle and aristocratic classes.[1] Women, though, saw the self-improvement opportunities opening up to their husbands and working sons, not least through Mechanics' Institutes, while their younger children benefited from the numerous Sunday schools, Industrial and Ragged schools that were springing up.[2]

Women of the middle classes – the wives and daughters of merchants, salesmen and the like – were increasingly educated by their fathers and husbands in order that they might one day take over the running of their businesses. But it was the boys who usually benefited most. They, like the scions of the aristocracy, might be sent to a boarding school or encouraged to join the officer classes of the British Army and the Royal Navy; the clergy too was an option.

The eighteenth century nevertheless saw a strengthening of the unhelpful (for women) conviction that a woman's place was firmly in the home while the man could do more or less what, when, where and with whom he so wished. Young, eligible ladies of the comfortable landed gentry were educated to a degree – in correct manners and conduct, polite conversation, appropriate dress, dancing, reading aloud, musical recitals, familiarity with a modern Romance language. Most, if not all, was designed to ensnare and reel in a good catch with as much of an allowance as possible coming in every year. Needless to say, the two English universities operating at the time – Oxford and Cambridge – were restricted to men, and wealthy Anglican men at that. Likewise, public schools such as Eton, Harrow, Westminster, Shrewsbury, Rugby and Dulwich were obviously men only.

But enlightenment was shining in from somewhere because from 1830 Mechanics' Institutes had opened their doors to women as well, while ladies' public schools such as Cheltenham Ladies' College in 1853, and Roedean School in 1885 were established by wealthy women who believed passionately that girls were intellectually just as capable as boys and so demanded an education to nurture and expand this.

The real turning point, however, came with the 1870 and 1880 Elementary Education Act which introduced government-funded mandatory education for all children between the ages of 5 and 10. This was often ignored by poorer rural parents who needed the money brought in by sending their children out to work in the fields. However, it did mean that all girls were guaranteed an education in law at least.

Teaching, especially in the early years, was taken up as a profession by many educated women. Some elected to be governesses, who lived in alongside household servants, others opened their own schools and took in pupils.

Something had been going right in the education of wealthy women, if the Blue Stockings Society is anything to go by. Elizabeth Montagu (née Robinson) was born in York in 1718 (d. 1800). She founded the Blue Stockings Society 'where literary topics were to be discussed, but politics, gossip and card-playing were barred'. Men's matters, then, were definitely not on the agenda.

The society was the creation of wealthy, educated, intelligent and independent-minded women rejecting the tedious post-prandial cigars and port ceremony which found their tribal men retiring to discuss how to prosecute the next war while their women were shepherded off to the parlour to gossip about the absent uninvited and speculate about their daughters' chances of acquiring the best possible match. We shall meet the Blue Stockings throughout the book.

Moreover, the Industrial Revolution and the relative independence it offered women were responsible for the proliferation of women's organisations predicated on education, empowerment, and social reform. Women's suffrage societies, temperance unions, and educational associations provided new platforms from which they could vocalise and disseminate their anxieties, advocate for their rights, and effect constructive and sustainable change in their communities. In so doing they advanced the status of women and paved the way for their advancement and progress during and beyond the Industrial Revolution.

Blue Stockings Society meetings apart, one of the places where women could meet and discuss issues of interest was the private members' club exclusively for women; these were springing up in London's clubland, previously monopolised by men in the Reform Club and Boodles, for example.

It is no accident that one of the activists who shared the stage and waved a banner at the Peterloo Massacre in 1819 was a woman – Mary Fildes. Of the 654 recorded casualties, at least 168 were women, four of whom died either at St Peter's Field or later from their wounds. We estimate that less than 12 per cent of the crowd were women, suggesting that they were at significantly greater risk of injury than men by a factor of almost three to one. Richard Carlile claimed that the women were especially targeted, a view apparently supported by the large number who suffered from wounds caused by military weapons.[3]

Chapter 2

Women in Industry and Commerce

We know that it was the men who provided the capital to fund the development and provision of the machinery which drove the Industrial Revolution in the textile factories, iron and steel works and down the mines. The men also set and paid the wages.

According to the Economic History Association, 'the invention of James Hargreaves' spinning jenny (1764), Richard Arkwright's water frame (1769) refers to two major inventions that significantly impacted the textile industry during the Industrial Revolution'. The spinning jenny enabled multiple spindles to spin yarn simultaneously, greatly increasing production.

> Richard Arkwright's water frame (1769), used water power to spin yarn and produced stronger, higher-quality thread suitable for warp. Britain began to manufacture cotton cloth ... Machines also appeared for other parts of the cloth-making process, the most important of which was Edmund Cartwright's power loom, which ... was widely used by the 1830s. While cotton was the most important textile of the Industrial Revolution, there were advances in machinery for silk, flax, and wool production as well.

However, we have seen how the emerging middle class was keen to guarantee a role for their wives and daughters in the many small businesses on the retirement or passing of the proprietor. These tables give an idea of the range of occupations active in 1851.

Table 1: Occupational distribution in the 1851 census of Great Britain

Occupation	Males (000s)	Females (000s)	% Female
Public Administration	64	3	4.5
Armed Forces	63	0	0.0
Professions	162	103	38.9
Domestic Services	193	1135	85.5
Commercial	91	0	0.0
Transportation & Communications	433	13	2.9
Agriculture	1788	229	11.4
Fishing	36	1	2.7
Mining	383	11	2.8
Metal Manufactures	536	36	6.3
Building & Construction	496	1	0.2
Wood & Furniture	152	8	5.0
Bricks, Cement, Pottery, Glass	75	15	16.7
Chemicals	42	4	8.7
Leather & Skins	55	5	8.3
Paper & Printing	62	16	20.5
Textiles	661	635	49.0
Clothing	418	491	54.0
Food, Drink, Lodging	348	53	13.2
Other	445	75	14.4
Total Occupied	6545	2832	30.2
Total Unoccupied	1060	5294	83.3

Source: B. R. Mitchell, *Abstract of British Historical Statistics*, Cambridge: Cambridge University Press, 1962, p. 60.

Census data would suggest that getting on for 30 per cent of the business population was female between 1851 and 1911, a figure higher than often given. Widows formed a substantial element of this when they elected to take the helm of their late husbands' businesses, having been involved in the running of the family business while their husbands were still alive.

Table 2: Factory workers in 1833: females as a percentage of the workforce

Industry	Ages 12 and under	Ages 13–20	Ages 21+	All Ages
Cotton	51.8	65.0	52.2	58.0
Wool	38.6	46.2	37.7	40.9
Flax	54.8	77.3	59.5	67.4
Silk	74.3	84.3	71.3	78.1
Lace	38.7	57.4	16.6	36.5
Potteries	38.1	46.9	27.1	29.4
Dyehouse	0.0	0.0	0.0	0.0
Glass	0.0	0.0	0.0	0.0
Paper	–	100.0	39.2	53.6
Whole Sample	52.8	66.4	48.0	56.8

Source: 'Report from Dr. James Mitchell to the Central Board of Commissioners, respecting the Returns made from the Factories, and the Results obtained from them.' *British Parliamentary Papers*, 1834 (167) XIX. Mitchell collected data from 82 cotton factories, 65 wool factories, 73 flax factories, 29 silk factories, 7 potteries, 11 lace factories, one dyehouse, one 'glass works', and two paper mills throughout Great Britain.

As noted, women were active in their own right with approximately 10 per cent of business owners listed in commercial directories being women.

Table 3: Business owners listed in commercial directories

Date	City	Male	Female	Unknown Gender	Percent Female
1788	Manchester	2,033	199	321	8.9
1824–5	Manchester	4,185	297	1,671	6.6
1846	Manchester	11,942	1,222	2,316	9.3
1850	Birmingham	15,054	2,020	1,677	11.8
1850	Derby	2,415	332	194	12.1

Sources: *Lewis's Manchester Directory* for 1788 (reprinted by Neil Richardson, Manchester, 1984); *Pigot and Dean's Directory for Manchester, Salford, &c.* for 1824–5 (Manchester 1825); *Slater's National Commercial Directory of Ireland* (Manchester, 1846); *Slater's Royal National and Commercial Directory* (Manchester, 1850)

Reproduced by the EHA: Joyce Burnette, 'Women Workers in the British Industrial Revolution'. EH.Net Encyclopedia, edited by Robert Whaples, 26 March 2008. https://eh.net/encyclopedia/women-workers-in-the-british-industrial-revolution

Table 4 shows us the trades occupied by women in Manchester:

Table 4: Women in trades in Manchester, 1788

Trade	Men	Women	Gender Unknown	Percent Female
Apothecary/Surgeon/Midwife	29	1	5	3.3
Attorney	39	0	3	0.0
Boot- and Shoe-makers	87	0	1	0.0
Butcher	33	1	1	2.9
Calenderer	31	4	5	11.4
Corn & Flour Dealer	45	4	5	8.2
Cotton Dealer	23	0	2	0.0
Draper, Mercer, Dealer of Cloth	46	15	19	24.6
Dyer	44	3	18	6.4
Fustian Cutter/Shearer	54	2	0	3.6
Grocers & Tea Dealers	91	16	12	15.0
Hairdresser & Peruke Maker	34	1	0	2.9
Hatter	45	3	4	6.3
Joiner	34	0	1	0.0
Liquor Dealer	30	4	14	11.8
Manufacturer, Cloth	257	4	118	1.5
Merchant	58	1	18	1.7
Public House/Inn/Tavern	126	13	2	9.4
School Master/Mistress	18	10	0	35.7
Shopkeeper	107	16	4	13.0
Tailor	59	0	1	0.0
Warehouse	64	0	14	0.0

Source: *Lewis's Manchester Directory for 1788* (reprinted by Neil Richardson, Manchester, 1984)

The city of York gives us two examples of female business owners: Mary Tuke (1695–1752) and Mary Craven (1826–1900) – two redoubtable women who steered their respective confectionery companies through difficult times to help make York the sugar, sweet and chocolate confectionery capital city it remains today.[6]

Guilds often jealously controlled access to trades, letting in only those who had served an apprenticeship, so earning the 'freedom' of the trade. Widowhood came to the rescue of many women who aspired to business ownership: we have observed how it was thought that in many cases the years of working alongside the late husband might often bestow upon the widow sufficient knowledge in the trade to be given the right to carry on the trade, bypassing the apprenticeship rule.[7] In the normal course of events, eighteenth-century women were apprenticed to a wide variety of trades, including butchery, bookbinding, brush making, carpentry, ropemaking and silversmithing.[8] The power of the guilds and the insistence on apprenticeships were, nevertheless, declining at this time.

Business management and ownership apart some women were able to compete at the highest levels of achievement in the industrial sector: a small number smashed through the glass ceiling restricting them from joining the higher echelons and actually owned factories. In Keighley, a town famous for its line of entrepreneurial and determined businesswomen, Ann Illingworth, Miss Rachael Leach, and Mrs Betty Hudson built and operated textile mills.[9] In 1833, Mrs Jess Doig Swanley (b. 1886) owned a power loom jute factory at Lochee, Angus employing sixty workers.[10]

The Victoria Press

Visionary Emily Faithfull set up her women-workers-only Victoria Press in 1860, a forerunner of the more recent Virago Press founded in 1973. The aims and scope of Victoria Press were to allow more women entry into the printing trade and to that end Faithfull hired and trained up women as compositors for her workshop. This predictably caused not a little hostility from the male-dominated London Typographical Society (then known as the London Society of Compositors); just as predictably the Union closed off compositors' work to all women using the feeble and paranoid justification that women lacked the mechanical ability and the intelligence to be compositors. Faithfull nevertheless persevered, and her Victoria Press thrived for years.

As well as publishing magazines including the popular weekly *Women and Work* and anthologies, the women at the Victoria Press were enthusiastic advocates for involvement in the workforce. Many of the pieces that the press published promoted women working in areas such as clock-making, hairdressing, and engraving, along with other hitherto male-dominated professions. Unsurprisingly, the press also published pieces promoting the absolute need for women to have educational opportunities equal to those offered to men and boys. In 1867, management of the press was sold by Faithfull to William Wilfred Head, a partner who continued to print articles advocating for the employment of women until 1882, even after buying Faithfull out in 1869.

The businesswoman

The Victoria Press is, of course, a good example of how women made a substantial and significant, though often underestimated, contribution to the commercial world and to local industry and services during and after the Industrial Revolution. Sometimes married women were engaged in trade independently from their husbands. Women most commonly ran shops and taverns, or worked as dressmakers and milliners, but, as we have seen, they appear in most of the trades listed in commercial directories. Manchester, for example, could boast six female blacksmiths and five female machine-makers in 1846. Between 1730 and 1800 there were 121 'rouping women' auctioning off estates in Edinburgh.

Piracy

Commerce on the wrong side side of the law but commerce all the same: Anne Bonny is the most celebrated women in the history of British piracy, plying her nefarious trade at the beginning of the Industrial Revolution around 1720. Her famous venture was the capture of the sloop *William* with thirteen or so fellow pirates including Mary Read. For the next two months they enjoyed a riot of piracy in the West Indies, attacking merchant shipping. All bad things come to an end and both Bonny and Read 'pleaded their bellies' at their trial in 1720; subsequent details are scarce but later records claim the two women were lesbian lovers.

Chapter 3

Women in the Professions

We usually regard the professionalisation of various occupations as a good and positive thing. Unfortunately, it has also resulted in the wholesale exclusion of women from work they had previously done, and done very professionally. Take medical care as an example – an occupation in which, since time immemorial, women rolled up their sleeves and got on with caring for the sick, the frail and the very young. Indeed, their role as midwives often extended beyond the modern iteration of this profession into the space now occupied by the obstetrician. But professionalisation of medicine in the early nineteenth century made maternity care, and other specialties where women operated at the fringes, an exclusively male occupation.

The Royal College of Physicians of London (RCP) was founded in 1518 by Henry VIII and, enduring male bastion as it was, believed that women were incapable of being doctors as evidenced, presumably, by the fact that the college only ever admitted graduates of Oxford and Cambridge; the intractable problem for women then was that these august institutions were just another two self-serving male bastions which at the time also only admitted men – a symptom of the insecurity felt by its board that women might outshine its male members? The RCP stricture on women meant that women had to be replaced at the end of the eighteenth century by men in that hitherto traditionally female profession. The term 'man-midwife', a(n) (oxy) moronic title, gained traction, flagging up a changing of gender roles.

Prominent man-midwives included Chamberlen, Manningham and Hunter and, according to the Royal College of Obstetrics and Gynaecologists (RCOG), 'man-midwives'

> came with the introduction of obstetric forceps, the use of which became the prerogative of doctors. As in many other walks of life at this time, the use of instruments and machinery was initially not thought to be suitable for women.

Soon into the nineteenth century the 'man-midwife' silliness fell into disuse, but women continued to be exempted from professionalisation, their role filled by physicians or surgeons in all matters obstetric.

The exclusion of women caused quite a stir amongst some, from as far away as the USA; this satirical letter to the *Journal of the American Medical Association* in 1896 begins as follows, advocating the early introduction of women to professional medicine in 'such an enlightened city as London':[11]

> The recent action of the Royal College of Physicians & Surgeons, England, in declining, by a narrow vote, to grant women their diplomas, is indicative both of the remaining prejudice against women doctors and the progress of their efforts for complete recognition by the profession ... Also the fact seems to have been overlooked that the 'new woman' is still comparatively new and has hardly had time to evolve that increase in brain and the desexing process which are finally to fit her with that keen, cool, deliberative mind requisite to rule the earth on scientific principles. The 'new woman' may not possess the charms of a Cleopatra, but she craves her power...

The London School of Medicine for Women

Some years before, in 1874, tangible progress had been made with the establishment of the London School of Medicine for Women (LSMW), established as the first medical school in Britain to train women as doctors. There was an obvious bias towards women's medicine as the aim was, in the words of Percy Edmunds,

> to provide educated women with the necessary facilities for learning and practicing midwifery and other branches of medicine while also promoting their future employment in the fields of midwifery and other fields of treatment for women and children.[12]

The school was a result of a collaboration between a group of passionate, pioneering women nurses and physicians: Florence Nightingale, Sophia Jex-Blake, Elizabeth Garrett Anderson, Emily Blackwell and Elizabeth Blackwell with Thomas Henry Huxley. Jex-Blake's exasperation at not being able to be awarded a medical degree (at a time when women were barred from British medical schools) gave the project added momentum: she was expelled from Edinburgh University for her troubles.[13] Other disaffected former colleagues joined her at the London school, including the irrepressible Isabel Thorne, who succeeded her as honorary secretary in 1877. Jex-Blake left to open a medical practice in Edinburgh from which she went on to found the Edinburgh School of Medicine for Women in 1886.

More enlightenment followed in 1877 (after the UK Medical Act of 1876 permitted medical authorities to license all qualified applicants irrespective of gender) when the Royal Free Hospital agreed to accept women students from LSMW to complete their clinical studies there, and by 1896 was known as the London Royal Free Hospital School of Medicine for Women, part of the University of London.

In 1894, the renowned Indian feminist Rukhmabai qualified in medicine after attending the London School of Medicine for Women. One result of this was that the number of Indian women students accordingly increased so that by 1920 the school opened a hostel for female Indian medical students. Later, in 1947, the school became co-educational by accepting men and was renamed the Royal Free Hospital School of Medicine.

The Edinburgh School of Medicine for Women

This was founded by Sophia Jex-Blake, with the help of the National Association for Promoting the Medical Education of Women.[14] Jex-Blake was appointed as both the director and the dean of the school. The inaugural class consisted of eight students, the youngest being 19 years old. Throughout its twelve years in operation, the school was challenged financially, not helped when the Edinburgh College of Medicine for Women was established by the Scottish Association for the Medical Education of Women. When the Scottish universities began to accept women medical students, as well as St Mungo's College and Queen Margaret College in Glasgow, the Edinburgh School of Medicine could no longer compete and closed in 1898, having trained eighty female students.

The Edinburgh Seven

The name has some significance: it is after Aeschylus's tragedy *The Seven Against Thebes*; other names include the Latin *Septem contra Edinam*; it involved many women over their four-year campaign.

After ten years of refusing to admit women into their medical school, some lecturers at the Extramural School in Edinburgh accepted women into their classes. Extramural medical education in Edinburgh began over 200 years before the university medical faculty was founded in 1726 and extramural teaching continued for a further 200 years after that.

The 'Edinburgh Seven' – Sophia Jex-Blake, Isabel Thorne, Edith Pechey, Matilda Chaplin, Helen Evans, Mary Anderson, and Emily Bovell – were the first women to be admitted into the medical programme in 1869, but things were never easy: the Seven had to pay higher fees than men because of their smaller class sizes. Professors could refuse to teach female students if they felt inclined so to do so these women had to arrange classes on their own instead of being assigned courses on the university programme. Astonishingly, despite having the same coursework and lectures, the all-female classes were graded quite differently from the male classes, so reducing the ability of these women to qualify for scholarships offered by the university based on good academic performance.

Trouble came also from the chaps at the school in a demonstration of breathtaking arrant sexism: this how the university officially described it:

> The male students at the university were displeased with the school's decision to admit females into the program. As a result, the male students exhibited aggressive behaviour and harassment toward their female counterparts on campus. Tension between the students reached a point of violence, and eventually a riot ensued. As the women arrived to take an anatomy examination, they were met by male student protestors who threw mud at them. The riot was publicized by local newspapers, which fostered public support for the plight faced by the female students. However, the violence had sparked fear at the university and resulted in the decision to allow women to participate in the medical program to be revisited. The university made the decision to revoke the women's admission, and they were declined from the program in the middle of their studies.

The Sex Disqualification (Removal) Act 1919 enabled women to join the professions for the first time.

Professionalisation of the clergy

Professionalisation of the clergy was also an effective means of excluding women. Adherents to the Church of England at the time obviously believed that their god had good reason for barring 50 per cent of his flock from aspiring to ordained clergy positions, leaving women to look after social work, education and teaching, and support for the poor and disadvantaged rather than aspiring to priesthood positions. However, those dissenters, the Methodists, had welcomed many women preachers during their early years, but even among the Methodists female preachers disappeared when lay preachers were replaced with a professional clergy in the early nineteenth century.[15]

Esther Hammerton (1711–1746)

One evening in February 1730, 18-year-old Esther, her father, her brother, and another man were digging a grave when her father dug too close to the pillars of St Mary's Chapel, Kingston-upon-Thames, which collapsed on top of them. Esther's father was killed, as was the other man, but she and her brother, though buried in the rubble for seven hours, were ultimately rescued and survived.

Esther took over the role of Sexton at All Saints Church from her mother in 1742, after her mother's death. This physical role involved digging graves and ringing church bells and the like. Women sextons were not rare at the time but Esther is remembered for helping to wipe out the stigma that many aimed at these women, considering their invaluable work 'unsuitable employment for a woman'.

Chapter 4

Votes for Women

The Industrial Revolution, and more particularly the freedom it afforded women outside the solitary home in factories where they communed all shift long with their peers, was a catalyst for early activism amongst working women. It focused their desire and demands for equality with their menfolk, for political rights, for much better working conditions and, most importantly, for the right to vote.

The franchise was extended after the First Reform Act of 1832 to take in more men, both in the Second Reform Act 1867 and the Third Reform Act 1884. About 58 per cent of the adult male population could vote by 1900, including some working-class men. But what about that huge percentage of the adult population that were women? Women were getting increasingly restless and disputatious.

Significant gains such as winning the rights to own property and to access higher education were important, but they were piecemeal. The big prize was always getting the elusive vote; it was acknowledged that without the vote, women would always be unequal, they would be at a disadvantage in many ways and that men would always exploit their political and social superiority over women.

And so by the 1860s women were organising themselves to establish parity with men. In 1866, a group of women organised a petition demanding the same political rights as men which they presented to Henry Fawcett and John Stuart Mill – two MPs supportive of universal suffrage.

Mill added an amendment to the 1867 Reform Act that would grant women the same political rights as men. Although it was defeated, by 196 votes to 73, it did lend impetus to the cause and led to the formation of the London National Society for Women's Suffrage (LNSWS). At about the same time it was responsible for the Manchester National Society for Women's Suffrage (MNSWS), whose meeting in the Free Trade Hall on

14 April 1868 offered a rare occasion for women to proclaim their rights and is widely regarded as the day when the women's suffrage campaign was born.

Other significant events which brought the vote closer for women before the momentous 1918 Act included the following:

- The Education Acts of 1870 and 1880 provided for mandatory primary school education for girls aged between 5 and 13.
- In 1880, women on the Isle of Man, a Crown Dependency, were granted suffrage, making it the first country to grant selected women the right to vote, specifically women who owned land. The IOM suffrage came thirty-seven years before the 1918 Act which extended (limited) suffrage to the UK. It was restricted to unmarried and widowed women aged 21 and over, who owned property.
- The Married Woman's Act of 1882 allowed women to own and control property in their own right for the first time:

A married woman shall, in accordance with the provisions of this Act, be capable of acquiring, holding, and disposing by will or otherwise, of any real or personal property as her separate property, in the same manner as if she were a feme sole [a woman without a husband, especially one that is divorced], without the intervention of any trustee.

- Both the Liberals and the Independent Labour Party (founded 1893) expressed some support for women's suffrage.
- In 1901 a group of women textile workers from northern counties presented a petition to Parliament with 29,000 signatures demanding the vote for women.
- The 1907 Qualification of Women Act allowed women to vote for borough councils.

The women textile workers' suffrage petition

By 1900, up to a quarter of a million women and girls were working in the mills in cotton-weaving towns like Preston, Blackburn and Burnley, where the majority of weavers were women. And they were taking home pay packets containing money not that far short of what the men took home. These were well-organised, articulate women – yet they all languished in a state of impotent disenfranchisement.

Esther Roper and Eva Gore-Booth were two empowered trailblazers who rose to the challenge of changing this anomalous situation. In Blackburn on May Day 1900, they launched a women textile workers' suffrage petition which by spring 1901 could boast over 29,000 signatures. Fifteen women cotton workers took the heavy roll of paper down to Westminster: their petition was so large it 'looked like a garden roller'.

Unsurprisingly, not much happened. In summer 1903, frustrated by the lack of progress on the suffrage front, Esther Roper, Eva Gore-Booth, Selina Cooper and two others decided to form the LWTWRC, the Lancashire Women Textile Workers' Representation Committee just before Emmeline Pankhurst set up the Women's Social and Political Union (WSPU). They were deeply critical of the by then famous Pankhurst militancy and violence, leaning noticeably towards the 1897 National Union of Women's Suffrage Societies (NUWSS), formed in 1897, popularly known and remembered as the suffragists and attracted by their non-confrontational and legal tactics with and a wide-ranging political agenda which put lobbying and petitioning above militant action.

The NUWSS was an expansive umbrella organisation formed in 1897 by a federation of seventeen societies to represent many local suffrage societies and was led by Millicent Garrett Fawcett. By 1913, the NUWSS embraced almost 500 regional suffrage societies with 54,000 members, for the most part educated and middle class; it can claim to be influential in the partial enfranchisement of women in the 1918 Representation of the People Act.

In April 1909, the NUWSS began to publish its own weekly newspaper, *The Common Cause* (1909–1920), edited by Helen Swanwick from an office in Manchester.[16] It was mainly financed by Margaret Ashton, who sold her house in leafy Didsbury to raise funds.[17]

Fawcett said in a speech in 1911 that their movement was 'like a glacier; slow moving but unstoppable'.

The London School of Economics (LSE) tells us that the offices of the WSPU were on the LSE campus. This was the London home of Emmeline and Frederick Pethick-Lawrence who also financed and edited *Votes for Women*, the newspaper of the WSPU.[18]

Chapter 5

Women and the Family

Industrialisation redefined the role of women, not only in the workplace but in the home as well. They became industrial wage earners with a significant contribution to the family income and, presumably more often than not, a say in how the combined earnings were budgeted and spent. But it was hard work.

We have already seen how working women were often expected to take on a double burden in their daily life: a gruelling twelve-hour shift followed very soon after by more hours at home taking care of husband and children, making the evening meal, keeping the house respectably tidy, and getting the children to bed. An estimated 750,000 factory working women were also mothers. Interestingly, one of the most common tombstone inscriptions honouring women in ancient Rome read *lanam fecit et domum servavit* – 'she worked the wool and did the housework' – not much change there then in 2,000 years.

What impact this daily separation of a mother from her children had it is hard to tell but the situation will have been ameliorated by the intervention of grandparents or familiar neighbours. However, parliamentary papers relating to northern industrial towns and cities reveal that some mothers enjoyed a legal and customary 'right' to take their children with them to work. Giving testimony before the 1876 Select Committee Inquiry into the Working of the Factories Act, Mr Bury, a schoolteacher, explained that it was a mother's right to have their children accompany them to the workplace and that this practice was a relic of old customs before the revolution.

We have records from 1865, in which Yorkshire mill mothers reveal that 'they had [their children] at the mill, and put them in a basket until [they] were ready to go home' for 'it was their "duty" to take' them, especially when 'they were at the breast'. Melanie Reynolds in *History Workshop* tells that:

> Inside the mills, women workers adapted the industrial architecture to ensure that they could supply breast milk to their infants and used the surface of the steam engine to warm their children's dinners of pies and potatoes. These working-class mothers avoided pap – an unsuitable mixture of water/milk, bread and sugar – 'like they shunned the devil'.[19]

This relaxed attitude to a primitive factory crèche system allowed women to escape the solitude of the domestic sphere and gave them personal independence, an identity and an opportunity to contribute to the family budget.

And it wasn't just factories: the *laissez faire* attitude extended to salt miners and chain, nail and brick workers. William Henry Edwards remembers his mother taking him to the brick yards in the Midlands:

> I had a little swing fixed up for me on the beam opposite the horse and round and round I would go all day. If the horse dropped anything, it had to be cleared away immediately to keep the path from becoming greasy and in bad condition. So, as I was travelled round and round, I kept a diligent eye for this, and was delighted when I could call out, 'Tom, Old Jack's messing again'.

Reynolds concludes:

> The industrial revolution did not mean a choice between maternal separation or confinement to the domestic sphere for all women. The workplace could be a place of both labour and childcare, so that mothers did not have to pay childcare costs and rely on others to bring their children up.

The mid-nineteenth century introduced the ideology of the male as the 'good provider', with women taking over most household and domestic tasks and in so doing imbued the concept with a kind of gendered normalcy. Thus, some women engaged primarily in domestic tasks such as childcare while the men went out to work. Being the sole wage earner in the family reinforced the man's traditional position as the head of the family.[20]

Childcare and opium

Childcare, its practicalities and cost have always been an intractable issue for working mothers and their families. It is disappointing, to say the least, that many of the issues relating to safe and affordable childcare during the Industrial Revolution and after still plague and restrict women, to a greater or lesser extent, today.

Women workers deployed a range of schemes and methods to care for their children. As we have just seen, sometimes childcare and work were compatible, and women took their children with them to the fields, factories or shops where they worked. Mary Collier's 1739 poem suggests that taking babies into the field was not uncommon; here is an excerpt:

> Our tender Babes into the Field we bear,
> And wrap them in our Cloaths to keep them warm,
> While round about we gather up the Corn;
> ... When Night comes on, unto our Home we go,
> Our Corn we carry, and our Infant too.
>
> —Mary Collier, *The Woman's Labour*,
> Augustan Reprint Society, 230, 1985, p. 10

Women who worked from home obviously in theory at least had more time to look after their offspring. Even then, some working mothers would sometimes give their infants a spoonful of mother's little helper – opiates such as 'Godfrey's Cordial' – in order to keep them sedated or a least sleepy. 'Godfrey's Cordial' (also euphemistically called Mother's Friend) and 'Dalby's Carminative' were two of the most widely used patent medicines administered to infants and children in England during the latter years of the eighteenth and early part of the nineteenth centuries.[21]

They each contained laudanum (tincture of opium) in a sweet syrup and were used mostly by mothers working in agriculture or industry: it ensured that she could work all the contracted hours of her employment, without being side-tracked by her infant, and thus preserved her contribution to the family income. Both branded preparations were used – almost always as self-medication without a doctor's advice – for a wide variety of symptoms ranging from social anxiety and colic to the severest forms of dehydration caused by explosive, bloody diarrhoea. Despite their innocuous names, they were sinister potions. Many infants died from opium poisoning in the early 1800s, not always as a result of their mothers' administering it but sometimes due to its surreptitious use by unscrupulous nurses and other child carers to keep infants in a deep state of sedation and thus of no bother to them.

Mary Colton, a lace worker from Nottingham, described her use of the drug to parliamentary investigators:

> Was confined of an illegitimate child in November, 1839. When the child was a week old she gave it a half teaspoonful of 'Godfrey's' twice a-day. She could not afford to pay for the nursing of the child, and so gave it 'Godfrey's' to keep it quiet, that she might not be interrupted at the lace piece; she gradually increased the quantity by a drop or two at a time until it reached a teaspoonful ... She has tried to break it off many times but cannot, for if she did, she should not have anything to eat.[22]

Sometimes children were too drugged up to eat and became malnourished as a result. Many died from starvation; others grew up sickly and stunted. Indeed, more babies died from starvation than from overdosing on opium itself. The Privy Council's Dr Greenhow stated in an investigation how children who were 'kept in a state of continued narcotism will be thereby disinclined for food, and be but imperfectly nourished'.[23] Marasmus, or inanition (exhaustion caused by lack of nourishment), and death from severe malnutrition were the consequence but the coroner would usually record these deaths as 'debility from birth', or 'lack of breast milk', or just 'starvation'. It took until 1868 before the Pharmacy Act restricted the sale of opium to legally registered pharmacists and chemists but there was no limit on amounts and the purchaser could buy anonymously. So drug abuse joined the non-registration of midwives and the adulteration of milk in the litany of shabby public health measures which could easily have delimited child mortality. Governments of the day passed the buck to the Pharmaceutical Society.

Factory work was tightly disciplined so women simply could not always bring their children into the work setting after that initial period in the very early days of the factory. Nevertheless, these constraints did not prevent women with small children from working. Working mothers deployed a number of other strategies to enable them to work: depending on older siblings, other relatives, neighbours, and, as noted, dame schools where children as young as two might be sent. Occasionally, young children would be left home alone, but the dangers in this were obvious so, as far as we know, only a few parents resorted to this. In areas where lace-making or straw-plaiting were prevalent, children were sent from about age 7 to 'schools' where they learnt their mothers' trade.

Mary Woodhouse, a midwife at the Manchester Lying-In Hospital, told an investigator in 1833 that factory women gave their babies 'the breast at breakfast, and at noon, and in the evening'.[24] Babies and children were left with childminders, while the women worked. Elizabeth Woodhouse, also a midwife, said the workers' children were left with 'any one they can get … some of them suffer very bad from this' and that 'many' factory women worked 'up to the very day … some up to the very hour' their baby arrived. Most mothers returned to work a fortnight after their confinement.

Medical experts protested against mothers who had worked in the mills since childhood because they developed narrow pelvises which was not conducive to childbirth; the night shift for pregnant women was also decried as it caused miscarriages, varicose veins and leg ulcers. All in all, the tiring work rendered the fatigued and undernourished pregnant woman unfit for motherhood. Mill owners and factory commissioners were in denial, as they were with regard to questions regarding the healthiness of babies born to these mothers.

In 1840, an observer of Spitalfields noted, 'In this neighbourhood, where the women as well as the men are employed in the manufacture of silk, many children are sent to small schools, not for instruction, but to be taken care of whilst their mothers are at work.'[25]

> At one straw-plaiting school in Hertfordshire, where the children start learning the trade at about 7 years old parents pay 3*d* a week for each child, and for this they are taught the trade and taught to read. The mistress has about 15 to 20 at work in a room; the parents get the profits of the children's labour.

Needless to say, the promised teaching and learning to read elements were somewhat thin on the ground: the children spent most of their time working.

One mistress complained that the children worked too much and learnt too little, 'In my judgement I think the mothers task the children too much; the mistress is obliged to make them perform it, otherwise they would put them to other schools.'

Here are examples from the Children's Employment Commission of mothers just getting on with it:

- Elizabeth Wells, who worked in a Leicester worsted factory, had five children, ages 10, 8, 6, 2, and four months. The eldest, a daughter, stayed home to tend the house and care for the infant. The second child worked, and the 6-year-old and 2-year-old were sent to 'an infant school'.
- In 1843 Mary Wright, an 'overlooker' in the rag-cutting room of a Buckinghamshire paper factory, had five children. The eldest worked in the rag-cutting room with her, the youngest was cared for at home, and the middle three were sent to a school; 'for taking care of an infant she pays 1*s* 6*d* a-week, and 3*d* a-week for the three others. They go to a school, where they are taken care of and taught to read.'
- Elizabeth Leadbeater, who worked for a Birmingham brass founder, worked while she was nursing and had her mother look after the infant.
- Mrs Smart, an agricultural worker from Calne, Wiltshire, reported, 'Sometimes I have had my mother, and sometimes my sister, to take care of the children, or I could not have gone out.'
- More often than not, though the childcare duties fell to older siblings – generally children of 9 or 10 years old, boys as well as girls. In a family from Presteigne, Wales, containing children aged 9, 7, 5, 3, and 1, we learn that 'The oldest children nurse the youngest.'[26]

There were the inevitable dreadful tragedies, such as one reported in *The Times* (6 February 1819):

> A shocking accident occurred at Llandudno, near Conway, on Tuesday night, during the absence of a miner and his wife, who had gone to attend a Methodist meeting, and locked the house door, leaving two children within; the house by some means took fire, and was, together with the unfortunate children, consumed to ashes; the eldest only four years old!

Childcare was expensive, prohibitively so for most. At the end of the eighteenth century, the price of childcare was about 1*s* a week, which was about 25 per cent of a woman's weekly earnings in agriculture.[27] In the 1840s mothers paid anywhere from *9d* to *2s 6d* per week for childcare, out of a wage of around *7s* per week.

In 1843 parliamentary investigator Alfred Austin reported, 'Where a girl is hired to take care of children, she is paid about 9d. a week, and has her food besides, which is a serious deduction from the wages of the woman at work. Who is paying for that?'[28]

Agricultural wages in the area were *8d* per day, so even without the cost of food, childcare costs were about 20 per cent of a woman's wage. One Scottish woman earned *7s* per week in a coal mine and paid *2s 6d*, or 36 per cent of her income, for the care of her children.[29]

Engels put the cost of childcare at *1s* or *18d* a week.[30]

Factory workers often made *7s* a week, so again these women may have paid around one-fifth or more of their earnings for childcare. The overseer of Wisbech, Cambridge, suggests a higher fraction; he reports, 'The earnings of the Wife we consider comparatively small, in cases where she has a large family to attend to; if she has one or two children, she has to pay half, or perhaps more of her earnings for a person to take care of them.'

Following years of campaigning and reports things started to improve in 1847 when the maximum working day in textile mills for women and those under 18 was reduced to ten hours. Factory inspectors interviewed workers soon after the new legislation was introduced, and met with the predictable mixed responses. One married weaver's weekly wage had fallen from *10s* to *8s 9d*, but she had 'more time' for her family and no longer paid her neighbours for 'cleaning the house, washing [and] cooking victuals'. Another woman, however, said that she would rather work twelve hours as she needed the extra money; her tailor husband was frequently redundant.

By 1873, over 26 per cent of women working in cotton mills nationally were married and 75 per cent of these were of childbearing age.[31] One factory inspector, Robert Baker, pointed out that, if these women were made redundant, they could not support their children or pay for any education. His solution to reduce the 'excess of infant mortality' in towns was to prevent women returning to work for several months after the birth of each child. But he failed to suggest how they could be supported or support in the meantime: statutory maternity benefit was not introduced until 1911.

On the positive side, some thrifty women helped themselves and their families by using penny savings banks. Sue Wilkes explains how 'At one Salford mill in 1874, which employed 500 women, workers paid in deposits ranging from 6*d* to 5*s* per week, getting 5 per cent interest on accounts over 10*s* in credit.' This factory bank system was 'a great boon to the married women … as a kind of lying-in club'. A wife put away money weekly for her baby's arrival 'unknown to her husband, and without his being able to meddle with it'. She no longer rushed back to work 'soon after her confinement', safeguarding her health and that of her baby. Sadly, despite all the benefits, only sixty-three of the 500 women mill workers had deposited savings.[32]

On the matter of thrift generally, by no means everyone was convinced. Although ideas of providence and hard work characterised middle-class families during the Industrial Revolution as described in Samuel Smiles' bestselling book *Self-Help*, Smiles (1812–1904), government reformer, counterbalances this with the sweeping assertion that the poorer classes were feckless and that their misery was 'voluntary and self-imposed – the results of idleness, thriftlessness, intemperance, and misconduct'.

Women also had to deal with the additional burden of society's demand to produce children to 'man' the factories, dig the coal and till the fields, as well as replenish the ranks of the army and crew the ships of the navy. Industrialisation led to a rapid increase in birth rates: pregnancy and childbirth clearly impact on the physical strength and stamina of women – it was not uncommon for families to have more than ten children.

Significantly, during the Industrial Revolution, there was a marked increase in the life expectancy of children. The percentage of the children born in London who died before the age of 5 decreased from 74.5 per cent in 1730–1749 to 31.8 per cent in 1810–1829. This may suggest an increase in relative wealth brought about by wage increases for men and women, improvements in health and hygiene and advances in family medicine and midwifery – particularly in relation to infection control.

Votes for women only became a reality in 1918 for women over age 30 courtesy of the Representation of the People Act, while the following year saw the first woman representative in Parliament in the shape of Nancy Astor, a wealthy American socialite married to an English lord. Americans, aristocracy, wealth and celebrity were in little evidence on the factory floor or at the local pit.

Chapter 6

Women in the Justice System

Women, as we know, got off to an unpromising start when it comes to English legal history. Women specifically make an appearance in only four chapters of the Magna Carta: 7, 8, 39 and 54. When the great work does actually mention women, it usually focuses on safeguarding the rights of wealthy widows and heiresses, particularly relating to property and marriage. While Chapter 8 protected widows from being forced to remarry, and Chapter 6 guaranteed widows their dower and inheritance, these mostly benefited the elite classes.

David Carpenter, translator of the Penguin edition of the *Magna Carta*, comments on the sorry situation around 1215 and for a very long time thereafter:

> If a free woman secured judgment of her peers (that is social equals), under its terms those peers would have been entirely male for women did not sit on juries. Chapter 39 also forgot about women altogether when it spoke of outlawry, for women were not outlawed they were 'waived', which meant left as a 'waif'. This had the same effect. A waived woman could be killed on sight just like an outlawed man. But the distinction showed how subject women were to men. Women took no oath of allegiance to the king because in theory they were always under the protection of a man – father, husband or lord. They were, therefore, never 'in law' and so could not be 'outlawed', hence they were 'waived' instead.

The Industrial Revolution saw little significant change in women's legal status; women had to wait until 1882 when the Married Women's Property Act gave women legal independence and control over their property and earnings. Generally though women's rights and independence remained limited, but married women fared worst since they were still considered under their husbands' control. Despite their mass invasion of the factory, mill and mine workspaces they were rewarded with lower pay and poorer working conditions compared to men. Laws usually favoured men and restricted women's rights and opportunities.

- Married women were 'feme covert', their legal existence was suspended during marriage, they being under the protection and control of their husband.
- Married women had limited control over their own property and income; their husband enjoying the right to manage and control it under the law.
- Divorce was complex and costly, and was stacked against women who often faced a loss of property and custody of children if they left their husbands without first obtaining a divorce.
- Much textile work, for example robotically operating simple machinery, was denigrated as 'women's work', while more skilled or higher-paying jobs went to the men.
- Women were often paid less than men for similar work, reinforcing the belief that their labour was of less value.
- Crime committed by women and girls was sometimes seen as less serious than male crime.
- Courts were often reluctant to convict mothers in cases of infanticide cases; witnesses were sometimes at pains to highlight maternal feelings before the birth to demonstrate a mother's lack of intent to kill.
- Women had limited access to legal representation and were often at a disadvantage in the male-orientated court system.
- Sir Mathew Hale did not help women's individual freedoms in 1736 when, in *The History of the Pleas of the Crown*, he made the following pronouncement, a pronouncement which was to be an albatross around the necks of English women for over two subsequent centuries: 'But the husband cannot be guilty of a rape committed by himself upon his lawful wife, for by their mutual matrimonial consent and contract the wife hath given up herself in this kind unto her husband which she cannot retract.'
- Neither did the 1824 Vagrancy Act which introduced the label 'common prostitute' to English Law, a pejorative term that was still used in the Street Offences Act 1959. The Vagrancy Act 1898 prohibited soliciting or importuning for immoral purposes and was at first intended as a measure in the fight against prostitution but in practice the legislation was almost exclusively used to convict men for gay sex. Finally, in 1995 it was ruled that 'common prostitute' only related to females and a man could not be charged with this offence. It was only with the Policing and Crime Act of 2009 that 'common prostitute' was replaced with 'person'.
- Before 1839, when a marriage broke down a mother had no automatic rights to her children. A truly desperate situation. The vexed issue had already

been debated in the Commons in 1837 when John Leader MP argued: 'As it stands at present, the law is entirely in favour of the husband and oppressive to the wife. A man who may be drunken, immoral, vicious, and utterly brutalized, may place his wife, who seeks to live separately from him, in this cruel dilemma – "You shall either continue to live with me, or you shall be deprived of your children." The wife, in such a case, has no redress.' The issue was taken up by women's rights campaigner Caroline Norton who had personal experience when she was herself denied access to her sons after her marriage broke down. She was instrumental in the change in law brought in by the Custody of Infants Act 1839. The new law allowed mothers to petition the courts for custody of their children.

- Soon after, the first Matrimonial Causes Act was enacted. Before 1857 you could only get a divorce by private act of Parliament. There were 314 of these between 1700 and 1857, most, unsurprisingly, initiated by the husband. How on earth could most women afford this? The Act transferred this function to the Court of Divorce and Matrimonial Causes. Later Acts would create a more equitable playing field for women in their attempts to file for divorce – a painfully slow process.
- The 1878 Matrimonial Causes Act was a legal landmark for those in violent marriages, enabling them to obtain separation orders.
- In 1937 grounds for divorce were extended beyond adultery to include desertion for over three years, cruelty and incurable insanity. And in 1969 the Divorce Reform Act enabled 'marital breakdown' to be used as a reason for seeking divorce.
- 'Coverture' in English common law dictated that until the first Married Women's Property Act was passed in 1870, after marrying a woman's rights to property were passed to her husband lock, stock and barrel. 'Thus, a woman, on marrying, relinquished her personal property – moveable property such as money, stocks, furniture, and livestock – to her husband's ownership; by law he was permitted to dispose of it at will at any time in the marriage and could even will it away at death'.
- The 1870 Act significantly ramped up the legal status of married women, granting them a greater degree of financial independence and control over their lives. It allowed women to legally own the wages that they earned, gifts and savings and to be able to inherit property, although all property in a woman's name before marriage still went to her husband. The Act also protected a woman's property from being seized by her husband's creditors. In 1882 the Women's Property Act granted women absolute control over their own money and property. In short, to do all these things as if they were single.

- The 1925 Law of Property Act finally allowed women to hold and dispose of property in the same way as men.

Although there are a number of early modern treatises dealing with the legal status of married women, such as *A Treatise of Feme Coverts, or the Lady's Law*, these were usually authored by men with the usual shortcomings. One of the first and only to be produced by a woman was Sarah Chapone's *The Hardships of the English Laws in Relation to Wives*, petitioning the king for legislative change. A long wait ensued.

Here are some examples of the gendered treatment women received in the eyes of the law:

- Before 1691, women convicted of stealing goods worth more than 10 shillings did not qualify for benefit of clergy. This was because this privilege, which was tantamount to a conditional pardon, was only originally applicable to the male profession of the clergy as the clergy admitted no women. Unlike men, such women were sentenced to death, though in reality they were often acquitted, convicted on reduced charges and sentenced to a lesser punishment, or even pardoned. The 'benefit of clergy' was a legal privilege where clergy, and later anyone who could read and claim to be a churchman, could be tried in a church court rather than a secular court for certain crimes. This meant they often faced less severe penalties, and sometimes escaped the death penalty, as church courts did not impose capital punishment. It originally aimed to protect clergy from secular courts and their harsh punishments. Imagine the abuse and corruption involved. The Church argued that allegedly miscreant clergy should be tried by their peers in ecclesiastical courts. It was finally abolished in England in 1827.
- Being convicted of treason or petty treason came at a heavy price for women: they were burned at the stake until 1790; men convicted likewise were hung drawn and quartered. The difference is probably accounted for by a reluctance to expose women's bodies in public; not much consolation there though.
- Women sentenced to death who successfully pleaded the belly – that they were pregnant – had their punishments respited, so that they could give birth, and often remitted completely. From 1848, reprieves granted to pregnant women were always permanent. 19 July 1917 saw *The Times* report:

> 'Pleading the belly' was a process by which a woman sentenced to death could request to be examined by a Jury of Matrons to establish whether

she was in the later stages of pregnancy. If she was determined to be 'quick with child' her execution date would be postponed until after she had given birth. The last recorded time a Jury of Matrons was employed in an English court was in 1917. This was rendered obsolete by the Sentence of Death (Expectant Mothers) Act 1931.

- The public flogging of women was abolished in 1817, having been in decline since the 1770s, while the public whipping of men persisted into the 1830s and was not abolished until 1862.
- Only men, of course, could be sentenced to enforced military or naval service.

Underlying all of this was the growing conviction that hard outdoor labour and military service were unsuitable for women and that women should focus on giving birth to and nurturing their children. There was also a palpable reluctance to physically punish women in public which had an influence on punishments handed down generally. No women were sentenced to the pillory after 1762, even if they were convicted of the same offences as men. Fewer women were selected for transportation than men because it was felt important to populate the colonies with those fittest for building their economies. This gendered selection, however, can have done little for building the next generation of colonists.

Crime committed by women was seen more and more as a manifestation of sexual deviance, rather than criminal.

Notes

1. www.oxford-royale.com/articles/history-womens-education-uk
2. Paul Chrystal, *Factory Girls and Climbing Boys: Women and Children at Work from the Industrial Revolution to 1900*, Pen & Sword, Barnsley, 2022.
3. M. L. Bush, *The Casualties of Peterloo*, Lancaster, 2005.
4. On the technology of the Industrial Revolution, see David Landes, *The Unbound Prometheus*, Cambridge University Press, 1969, and Joel Mokyr, *The Lever of Riches*, Oxford University Press, 1990.
5. Joyce Burnette, 'Women Workers in the British Industrial Revolution', EH.Net Encyclopedia, edited by Robert Whaples. 26 March 2008. https://eh.net/encyclopedia/women-workers-in-the-british-industrial-revolution
6. Paul Chrystal, *Rowntrees: The Early Years*, 2021; Paul Chrystal, *A History of Sweets*, Pen & Sword, Barnsley, 2021.
7. For one woman's struggle against the Guilds, see the trials and tribulations endured by Mary Tuke.

8. See K. D. M. Snell, *Annals of the Labouring Poor*, Cambridge University Press, 1985, Table 6.1.
9. See Francois Crouzet, *The First Industrialists*, Cambridge University Press, 1985, and M. L. Baumber, *From Revival to Regency: A History of Keighley and Haworth, 1740–1820*, Crabtree Ltd., Keighley, 1983.
10. First Report of the Central Board of His Majesty's Commissioners for inquiry into the Employment of Children in Factories, with Minutes of Evidence, British Parliamentary Papers, 1833 (450) XX, A1, p. 120.
11. www.genesontheweb.co.uk/g6/p6871.htm
12. J. S. C., 'The Royal Colleges and the Women', *JAMA*, 1896, XXVI(4): 189. doi:10.1001/jama.1896.02430560041011
13. Percy Edmunds (1911), 'The Origin of the London School of Medicine for Women', *The British Medical Journal*, 1 (2620): 659–660. doi:10.1136/bmj.1.2620.659-b.
14. Historic England, 'Former London School of Medicine for Women', historicengland.org.uk
15. J. M. Somerville, 'Dr Sophia Jex-Blake and the Edinburgh School of Medicine for Women, 1886–1898' (PDF), *The Journal of the Royal College of Physicians of Edinburgh*, 35(3): 261–7, 2005.
16. See Deborah Valenze, *Prophetic Sons and Daughters: Female Preaching and Popular Religion in Industrial England*, Princeton University Press, 1985.
17. *John Simpkin*, 'The Common Cause', *Spartacus Educational.*
18. 'Margaret Ashton', *Suffrage Pioneers 1918–2018.*
19. www.lse.ac.uk/library/collection-highlights/womens-suffrage#
20. *Labour and Love: A Herstory of Work and Childcare in the Industrial Revolution*, April 2018. www.historyworkshop.org.uk/womens-history/labour-and-love
21. Shawn L. Christiansen, 2001, 'Why the "Good Provider" Role Still Matters: Providing as a Form of Paternal Involvement', *Journal of Family Issues*, 22(1), 84–106.
22. Children's Employment Commission: Second Report of the Commissioners (Trades and Manufactures), British Parliamentary Papers, 1843 (431), Vol. XIV, p. 630.
23. See Thomas E. Jordan, 'The Keys of Paradise: Godfrey's Cordial and Children in Victorian Britain', *Journal of the Royal Society of Health*, 107(1), 1987, pp. 19–22; Terry M. Parssinen, *Secret Passions, Secret Remedies: Narcotic Drugs in British Society, 1820–1930*, Manchester, 1983; Elizabeth Lomax, 'The Uses and Abuses of Opiates in Nineteenth-Century England', *Bulletin of the History of Medicine*, 47(2), 1973, pp. 167–176; Paul Chrystal, *Factory Girls and Climbing Boys: Women and Children at Work from the Industrial Revolution to 1900*, Pen & Sword, Barnsley, 2022 (Laudanum).
24. Anthony S. Wohl, 'Opium and Infant Mortality', 1989, https://victorianweb.org/science/health/health4.html
25. Reports from Assistant Handloom-Weavers' Commissioners, British Parliamentary Papers, 1840 (43), Vol. XXIII, p. 261.
26. Sue Wilkes, 'A Working Woman's Place', *History Today*, 67(6), June 2017.
27. Eden, 1797, vol. iii, p. 904.
28. Reports of Special Assistant Poor Law Commissioners on the Employment of Women and Children in Agriculture, British Parliamentary Papers, 1843 (510), Vol. XII, p. 26.
29. British Parliamentary Papers, 1844 (592), Vol. XVI, p. 6.
30. Friedrich Engels, *The Condition of the Working Class in England*, 1926 (1845), p. 143.
31. Sue Wilkes, 'A Working Woman's Place'.
32. Ibid.

Part Two

The 200 Trailblazers

Engineering

Given the educational barriers excluding women at the time, engineering is one of the professions which we would expect men to dominate during the Industrial Revolution; Matthew Boulton and James Watt are two outstanding examples with modifications of the steam engine as invented by Thomas Savery in 1698. Boulton and Watt's work laid down the very bedrock of the revolution. Hargreaves and Cartwright were two others, developing the 'Spinning Jenny' (1764) and the power loom (1785) respectively. Engineers were also busy with the development of the emerging railway industry as well and in the mines.

Here are three women who proved by their achievements that engineering was not a male preserve and that women too could excel in this rapidly developing field.

Sarah Guppy (1770–1852)

Guppy was the first woman to patent a bridge, which she did in 1811. Famous also for developing a number of other domestic and marine products, Guppy was by nature self-deprecating and modest, writing at one point in her professional life: 'It is unpleasant to speak of oneself – it may seem boastful, particularly in a woman.'

After an incorrect entry in the *Oxford Dictionary of National Biography* in 2016, now corrected, Guppy was wrongly credited with the design of Brunel's Clifton Suspension Bridge (1864). She had patented her design for a chain bridge in 1811 which came to nothing. Brunel's winning design differed from Guppy's patent in a number of significant ways: it had a deck suspended from flat wrought-iron bar links rather than resting on top of chains like Guppy's; and it did not feature foundations on the river bed which were a key component of Guppy's design as it was constructed on rock above the high tide level to avoid damage from water erosion.

Guppy also obtained patents in the early nineteenth century for the following intriguing inventions:

- a method of keeping ships barnacle-free, giving her a Royal Navy contract worth £40,000
- a bed with built-in exercise equipment in 1831; the drawers under the bed could be used as steps for exercise, while bars suspended from the ceiling provided for upper-body strength training
- a tea or coffee urn which could cook eggs in the steam complete with a small dish to keep toast warm – Wallace and Gromit, nineteenth-century style
- a device for 'improvements in caulking ships, boats and other vessels'
- the fire hood or Cook's Comforter
- a novel type of candlestick that enabled candles to burn longer

We have seen how women in the Industrial Revolution were forbidden to own property in their own name. This included patents, which were seen to be intellectual property and had some value, so Guppy made her patents in her husband's name. In 1811 she patented a 'New Mode of Constructing and Erecting Bridges and Railroads without Arches or Sterlings whereby the Danger of being washed away by Floods is Avoidable' – a method using piling for the construction of a suspension bridge.

> I do fix or drive a row of piles, with suitable framing to connect them together, and behind these I do fix, or drive, and connect, other piles or rows of piles and suitable framing, or otherwise, upon the banks of the said river or place.

This was some seven years before Thomas Telford started on the Menai Suspension Bridge. When Telford asked her for permission to use her patented design for suspension bridges, Guppy gave Telford permission to use her design and ideas for free, but the records show scant acknowledgement of her remarkable achievements.

Nevertheless, as a friend of Brunel and his family, Guppy became involved in the Great Western Railway, writing to the directors with ideas and giving her support. In 1841 she wrote a particularly important letter recommending the planting of willows and poplars to stabilise embankments; she also contributed significantly to the design of SS *Great Western* and SS *Great Britain*.

Until recently, history had done a good job burying Guppy's advances in bridge engineering: her contributions have clearly been hijacked by prominent male engineers. At the same time, history has virtually ignored her prominent role in the social reforms of her day, and her philanthropic contributions to education, especially the education of girls and women; she released pamphlets on public health, agriculture, education and welfare of animals, all in effect a move towards a fairer society.

Sarah Guppy was selfless and generous, and cared more for the greater good than for her own personal gain or kudos.

Later in life she authored *The Cottagers and Labourers Friend* and the two-volume *Mrs Guppy's Dialogues for Children* (1800); profits went to a charity school for girls in Bristol.

As evidence of her voracious appetite for knowledge and learning, records show that Guppy borrowed more books from the Bristol Library Society than any other female borrower in the years up to 1811.

Henrietta Vansittart (1833–1853)

Self-trained Vansittart was awarded a patent for a screw propeller named the Lowe-Vansittart propeller. Her speciality was ship propulsion.

Shortly after marrying, Vansittart began to study her father's work on ship propulsion. She had sailed with her father on HMS *Bullfinch* to test their latest version of his screw propeller, which ignited her interest, in 1857. This Lowe propeller had already been fixed to many British warships, but James Lowe saw little financial reward because of infringement battles. Vansittart's father died in 1866, having been killed by a wagon on Blackfriars Road in London; she began working and experimenting at pace, presumably hoping to gain the recognition which eluded her father.

In 1868, Vansittart was awarded a patent (no. 2877), for an improvement on her father's work – what she called the Lowe-Vansittart propeller: this showed 'improvements in the construction of screw propellers', resulting in ships travelling faster, with fewer vibrations, and using less fuel. It was subsequently fitted on over thirty ships, including HMS *Druid*, the *Scandinavian*, and the SS *Lusitania*. The Maritime Archaeology Trust tells how:

> the iron hulled passenger/cargo steamer SS *Scandinavian* underwent testing using the original two-cylinder double acting 400hp engine with the Lowe-Vansittart propeller. The Ships Master and the Liverpool Pilot witnessed first-hand how this device used 15% less of the fuel while increasing the speed of the boat of ¼ knots. In addition, there were no vibrations, a tighter turning circle, and, compared with other contemporary propellers, the Lowe-Vansittart propeller had no problems steaming in reverse, for example when leaving the dockside.

In the 1860s, Henrietta improved the design with curved rather than straight blades for greater efficiency. Henrietta travelled to Tynemouth for the 1882 North East Coast Exhibition of Naval Architecture and Marine Engineering to demonstrate her propeller. Sadly, she was found in a state of confusion and, on the day her patent expired as fees had not been paid, she was admitted to Tyne City Lunatic Asylum at Gosforth, where she died on 8 February 1883.

Laura Annie Willson MBE (1877–1942)

Willson started working life at the age of 10 as a 'half-timer' in a local Halifax textile factory. Half time working came in to allow children to labour for half a day and then spend the rest of the time at school, often within the factory.

A notable English engineer and suffragette, who was twice imprisoned for her political activism, she was one of the founding members of the Women's Engineering Society and was the first female member of the Federation of House Builders.

Business and Industry

Mary Tuke (1695–1752)

The Tuke tea and coffee business originated with Mary Tuke, a resolute Quaker woman; it was the Tuke business that the Rowntree family bought and turned into one of the world's biggest and most successful confectionery companies. Rowntree's still trades today in York, albeit as part of the Nestlé conglomerate. Without Mary, the world would be a poorer place, what with neither Polos nor KitKats.

Mary was an ambitious woman and was determined to establish her grocery business in 1725 York; no easy challenge, burdened as she was being a woman and a Quaker woman at that. She duly became embroiled in a series of tortuous commercial legal wrangles with the male-dominated York Merchant Adventurers Company. To trade in York you had to be a freeman of the city: to acquire such a status it was necessary to pay £25, to serve an apprenticeship, or else to be related to an existing freeman. Mary did win the status of freeman by patrimony, citing her father as a deceased member. The inscription on the *Freeman's Roll of the City of York* reads, in a strange fusion of English and Latin: 'Maria Tewk, spinster Fil Willelmi Tuke, blacksmith.'

Unfortunately, further hurdles were placed in her way: she was still debarred from trading because she was now required to be a member of the Merchant Adventurers Company, or else be granted a licence by them; but she had no means of achieving either. Flouting what was essentially a pointless and outmoded law, Mary was prosecuted for trading without a licence at the Midsummer Court in June 1725 ('Merchandising and following Trade without being free of this Fellowship'). In July 1728, the Merchant Adventurers finally relented and allowed her to trade at the pleasure of the court, on payment of five shillings every six months, and on condition she bought all her goods locally and took on no apprentices. In 1732, Mary was allowed to trade in perpetuity after a one-off payment of £10. The significance of Mary's brave victory over this antiquated but still powerful monopoly should not be underestimated; her tenacity and courage served as a great example to the Tukes, and others – not least the Rowntrees – who eventually bought her business outright.

In 1733, Mary married Henry Frankland, a local Quaker stuff weaver. Henry changed his trade from weaver to grocer in 1736; Mary was left on her own in 1739 when Henry died.

William Tuke III, or 'Old William' (1732–1822), inherited the business on Mary's death in 1752, aged 20 and with two years to run on his apprenticeship. The Castlegate shop now specialised in the sale of tea, coffee, chicory and drinking chocolate; William's son Henry (1755–1814), at that time aspiring to become a doctor, joined the firm in 1770; after years of studying in his spare time he gave up his 'taste for the physic' in 1785 in a typically Quakerish act and committed himself to the business as a partner. Henry Tuke was keen to expand his product base: possibly with an eye on Fry's success in Bristol, he branched out to manufacture Tuke's own cocoa and chocolate in 1785.

Henry's son, Samuel (1784–1857), joined the firm in 1795, became a partner in 1805 and managed it until 1852.

Samuel was a good friend of Joseph Rowntree Sr. (1801–1859), as well as being related to him through marriage. Samuel and Joseph Rowntree had much in common apart from their Quaker faith: campaigning for the return of William Wilberforce, supporting The Retreat, setting up the Friends' Provident Association in 1832 and establishing famous schools which would later become Bootham and The Mount, both still educating today.

And so it was in 1862 that Henry Isaac Rowntree bought the Tukes' cocoa and chocolate business, calling it the 'Cocoa, Chocolate and Chicory Works'. The stimulus may have come from the fact that Joseph Rowntree's existing business in Pavement was already well established and capably run by Henry's brothers, John Stephenson and Joseph II (1836–1925) who were made partners on reaching their majority; it may have been triggered by a desire to provide a non-alcoholic alternative beverage in keeping with Quaker temperance as pioneered by Joseph Fry in Bristol and then George Cadbury in Bournville. Lewis Fry and George Cadbury were themselves apprenticed at the Pavement shop. Whatever, Henry obviously benefited financially from his father's will and perhaps he chose the Tuke option – and the independence it offered – rather than a stake in the Pavement grocery business and the constraints that would necessarily go with it. He may even have identified a growing market, indeed the birth of a mass market: in 1820, cocoa consumption in Britain was 267,000 lbs (0.01 lb per head of population); this had grown to 4,583,000 (0.16 lb per head) by 1860 and by 1900 to 43,680,000 lb or 1.06 lb per head.

This and the global success which followed for the next 150 years (so far) can be attributed to the determination of Mary Tuke.

Frances Pawlett (1720–1808)

It is not permitted to manufacture Stilton in Stilton since it lies outside the designated geographical area, or Geographical Indication (GI). And Frances Pawlett did not, as we are often led to believe, 'invent' Stilton cheese. But anyone who has a taste for fine cheeses has a debt of gratitude to Frances for it was she who commercialised and standardised the shape and weight of that king of the post-prandial Stilton cheese. Working from the family farm at Wymondham in Leicestershire, she initially supplied the Bell Inn, run by her brother-in-law Cooper Thornhill, in Stilton. The pub can still be seen just off the A1 going north.

Daniel Defoe laid it on a bit thick in 1724 when he wrote:

> We pass'd Stilton, a town famous for cheese, which is call'd our English Parmesan, and is brought to table with the mites or maggots round it, so thick, that they bring a spoon with them for you to eat the mites with, as you do the cheese.[1]

Note

1. Daniel Defoe, *A Tour thro' the Whole Island of Great Britain* (1724–1727).

Mary Ann Craven (1826–1900)

Given York's status as one of, if not *the*, world's confectionery capitals, it should come as no surprise to find another trailblazer in Mary Craven (née Hick) from the city. Mary ran two businesses and established one of the world's biggest producers of boiled sweets; Craven's dates back to 1803 and still trades today in York under the name of Tangerine Confectionery.

Mary Ann was born in 1826. In 1833, Thomas Craven arrived in York aged 16 as apprentice to Thomas Hide, his brother-in-law; he bought the right to trade in York when he became a Freeman in 1840; and in 1851 he married Mary Ann Hick – she was 22, he 34. In 1860, Mary Anne's father, Joseph Hick, himself a confectioner with a business, died; then in 1862 Mary Ann's husband died, leaving her with three young children (one boy and two girls aged 8, 6 and 4) to raise and two businesses to run. Mary Ann Craven was equal to the challenge: she amalgamated the two businesses, changed the name of the company to M. A. Craven, and ran it until 1900.

Toffees and humbugs and other sugar confectionery became global bestsellers – and remain so. Staff numbers had increased to 800 by 1908 – a sizeable business by York standards. There were four Craven's retail shops in the city one of which, Craven's Mary Ann Sweet Shop, was in the Shambles and featured a sweet museum on the first floor where visitors could see 150 years of the 'Art, Trade, Mystery and Business of the Confectioner'.

In 1920, the Coppergate factory was named the French Almond Works in recognition of the importance of that product line. In 2008, Tangerine Confectionery bought Monkhill from Cadbury Schweppes for £58 million.

Beatrix Potter (1866–1943)

Beatrix Potter, of course, is embedded in many a memory as the author of the famous series of twenty-three children's books, beginning in 1902 with the self-published *The Tale of Peter Rabbit*, all beautifully written and illustrated. Two hundred and fifty million-plus sales later we must set aside those memories and focus on Beatrix's skills as a prominent entrepreneur and merchandiser.[1] In 1903, when Marie Curie was being awarded the Nobel Prize for Physics, Peter Rabbit became the first fictional character to be made into a patented stuffed toy, making him the oldest licensed character. The juxtaposition of the two women's trailblazing achievements is striking: one has done inestimably valuable work in diagnostic and therapeutic medicine, the other has generated endless pleasure for generations of children (and their parents) across the globe.

What are the significant childhood and adolescent factors which influenced Beatrix Potter?

She was born into a comfortable middle-class family from the Manchester area, moving to West Brompton, London, where Helen Beatrix was born; she was educated by three governesses, largely isolated from other children. Linda Lear (2007 and 2008) explains how the last of whom was Annie Moore (née Carter), just three years older than Potter, who tutored Potter in German as well as acting as lady's companion. She and Potter remained friends throughout their lives, and Annie's eight children were the recipients of many of Potter's picture letters. Significantly for us and for Beatrix it was Annie who later suggested that these letters might make good children's books.

She kept many pets and spent her holidays in Perthshire and the Lake District, initially at Wray Castle near Windermere, developing a love of landscape, flora and fauna, all of which she closely observed and painted. Potter's study and watercolour paintings of fungi led to her being widely respected in the field of mycology.

Predictably, sexism and the defence of male bastions were never far away: rebuffed by William Thiselton-Dyer, the Director at Kew, because of her sex and amateur status, Potter wrote up her conclusions and submitted a paper, *On the Germination of the Spores of the Agaricineae*, to the Linnean Society in 1897. It was introduced by George Massee because, as a woman, Potter was able neither to attend proceedings nor read her paper. She subsequently withdrew it, realising that some of her samples were contaminated, but continued her microscopic studies for several more years. In 1997, the Linnean Society issued a posthumous apology to Potter for the sexism it displayed in its handling of her research.

She became that successful children's author we all know and used her royalties and a legacy to buy Hill Top Farm in Near Sawrey. She bought up additional farms to restore and preserve the unique hill country landscape.

In 1913, at the age of 47, she married William Heelis (1871–1945), a local solicitor with an office in Hawkshead. Potter was a prize-winning breeder of Herdwick sheep and a prosperous farmer keenly interested in land preservation and the fell farming way of life.

Potter died on 22 December 1943 at Near Sawrey, leaving most of her property to the National Trust, and is accordingly credited with preserving much of the land that now is the beautiful Lake District National Park.

Note

1. Potter wrote sixty books in total.

The Sciences

Lady Henrietta Antonia Clive, Countess of Powis (1758–1830)

Henrietta Clive is remembered for her trailblazing work in botany, geology and as a writer, not least as one of the pioneers of women's travel writing. Botany and geology, of course, are not obvious or easy career choices for women due to the values of the day and the consequent barriers placed in the way of women to obstruct any progress they may make in these fields. In botany we also have Mary Pirie; in geology Mary Anning – both covered below.

Much of her inspiration and aptitude for both disciplines was generated by her days in India while her husband, the grasping Clive of India, was Governor of Madras (1798–1803). Henrietta began collecting in India and was apparently the first aristocratic woman to collect minerals as a hobby.[1]

Lady Clive used her connections in the mineral collecting and dealing world to contact prominent collectors and mineral dealers, such as mineralogist and botanist James Sowerby, John MacCulloch and Louisa Finch, the Countess of Aylesford, naturalist and botanical illustrator.[2]

The minerals in Lady Clive's collection, numbering up to 1,000, are arranged systematically by chemistry, usual practice in the early nineteenth century. In 1817, she organised her collection in two handwritten catalogues: *Volume 1 – Earthy Minerals* and *Volume 2 – Metallic Minerals*. She used a numbering system to identify each specimen, with small numbered labels affixed to the specimens, ensuring the collection remained remarkably complete to this day – in short she has given us one of the earliest systematic mineral collections in Britain.

Lady Clive also created a garden and kept a record of the plants around Mysore and the Carnatic region.

Her journals are one of the first written accounts of India by a British woman. Published in the 2010 edited collection *Birds of Passage: Henrietta Clive's Travels in South India 1798–1801*, they are an important milestone in the emergence of female travel writers and their elevation to the level of their male counterparts.

Notes

1. The Fabulous Mineral Collection of Lady Henrietta Antonia Herbert, Countess of Powis'. Williams, Sally. Laid bare the passions and obsessions of Lady Henrietta, a nineteenth-century rock chick Wales Online 2010.
2. After settling in Warwickshire, the Countess of Aylesford started studying the region's flora. She produced over 2,800 botanical watercolour drawings and was a correspondent of botanists such as William Withering, William Thomas Bree, and George Don. She also documented about 30 first records of plants from Warwickshire. She amassed an extensive collection of minerals, which was acquired by Henry Heuland after her death. Her plants are collected in Oxford University, and her minerals and manuscripts in the Natural History Museum. John MacCulloch was the first geologist to be employed by the British government in Britain and is best known for his pioneering texts on geology and for producing the first geological maps of Scotland. He introduced the word 'malaria' into the English language.

Etheldred Benett (1775–1845)

In the nineteenth century, geology and palaeontology were not noted for welcoming women into their male-dominated midst; in fact, women were not permitted to join the Geological Society of London until 1904 because… they were not men, whatever their scientific credentials and achievement in the field. Today, we learn from the same society that the geologist and botanist Aylmer Bourke Lambert, her sister-in-law's half-brother, encouraged her and her sister Anna Maria Benett to study natural history. Whilst her sister concentrated on botany, Benett took up the newly fashionable study of fossils.

By 1809, Benett had begun to acquire a significant collection of material. Her independent wealth (she never married) meant that she was able to collect high-quality specimens from the many working quarries in the area, as well as from her holidays to the Dorset coast. Such was the importance of her collection that it became the first port of call for geologists studying the Wiltshire area. She also sent duplicate specimens to museums all over the country (including the Geological Society) and published books on her collection.[1]

Apparently her unusual name and pioneering achievements in what was perceived to be the masculine science of geology meant that she was, unsurprisingly, regularly mistaken for a man. In 1836, the Natural History

Society of Moscow made her a member but the diploma was ascribed to 'Dominum [Master] Etheldredus Benett'.

Notwithstanding, her fossil collection, considered one of the largest at the time, played a part in the development of geology as a science. R. J. Cleevely confirms:

> Benett's speciality was the Middle Cretaceous Upper Greensand in Wiltshire's Vale of Wardour. Her collection was one of the largest and most diverse of its time, resulting in many visitors to her home. The collection consisted of over 1500 specimens that are now being recognised due to the resurfacing of said collection by a publication in 1989. Some fossils in her collection were the first to be illustrated and described, whilst others were extremely rare or unusually well preserved.[2]

Notes

1. www.geolsoc.org.uk/the-library/online-exhibitions/Women-and-Geology/Etheldred-Benett
2. R. J. Cleevely, 'Miss Etheldred Benett (1775–1845): A Preliminary Note on her Correspondence', *Wiltshire Archaeological and Natural History Magazine*, 97, 25–34, 2004.

Mary Fairfax Somerville (1780–1872)

Mary Fairfax Somerville was a trailblazing Scottish student of mathematics and astronomy, overcoming the usual limited educational opportunities encountered by all the clever women in this book by teaching herself (often in a clandestine way), after her parents decreed that studying damaged women's health and tried to stop her learning any more mathematics. In her *Personal Recollections*, Somerville notes that the boys learned Latin at the village school, while 'it was thought sufficient for the girls to be able to read the Bible; very few even learnt writing'.

Her uncle, Dr Thomas Somerville, assured her that in earlier times many women had become elegant scholars. Mary Brück then goes on to inform us that he then proceeded to enable her to improve her Latin by reading Virgil with her. On the other hand while staying with another uncle, William Charters, in Edinburgh, Somerville was sent to a dancing school, where she learned manners and how to curtsey.

She went on to translate Pierre-Simon Laplace's *Mécanique Céleste* into a popular English edition, and write bestselling textbooks on science and geography. Mary was, in her own words, 'allowed to grow up a wild creature'. In 1835, she became one of the two first female members of the Royal Astronomical Society. Somerville College was named after Mary Somerville in 1879, reflecting its freethinking and liberal ethos.

We hear of the liberating impact the death of a husband can have on a number of our talented trailblazers. Mary is a good example; she found that widowhood and a comfortable inheritance had left her both emotionally and financially independent; she was free to study and mastered *Ferguson's Astronomy* and became a student of Isaac Newton's *Principia*, despite opposition from family and friends.

In 1826 she presented her paper entitled 'The Magnetic Properties of the Violet Rays of the Solar Spectrum' to the Royal Society. Apart from the astronomical observations of Caroline Herschel, this was the first paper by a woman to be read to the Royal Society and published in its Philosophical Transactions; it distinguished her as a skilled scientific writer respected among her fellow scholars.[1]

Shane Wood (Agnes Scott College, Atlanta, GA) lists Mary's subsequent high achievements:

- The Mechanism of the Heavens was a great success, probably the most famous of her mathematical writings. In recognition, a portrait bust of her was commissioned by her admirers in the Royal Society and placed in their great hall.
- 1834 saw the publication of *The Connection of the Physical Sciences*.
- She and Caroline Herschel were elected in 1835 to the Royal Astronomical Society, the first women to receive such an honour. She was given a pension of £200 per year from the king.
- In 1848, Mary published *Physical Geography*, a work for which she was preached against in York Minster for her troubles; it proved her most successful yet and was widely used in schools and universities for the next fifty years.[2]

Notes

1. Louise S. Grinstein and Paul J. Campbell, *Women of Mathematics: A Bio-Bibliographic Sourcebook*, London, 1987, p. 213.
2. Grinstein and Campbell, 1987, p. 214; https://mathwomen.agnesscott.org/women/somer.htm

Mary Anning (1799–1847)

Mary Anning was another such excluded woman, even though she became internationally acclaimed for her landmark discoveries in Jurassic marine fossil beds in the Blue Lias cliffs at Lyme Regis. Anning's findings helped change the way we thought about prehistoric life and the history of the Earth, no less.

As a teenager, her impressive discoveries included:

- The first correctly identified ichthyosaur skeleton when she was 12 years old in 1811.
- The first two virtually complete plesiosaur skeletons; one sold for £200.
- The first pterosaur skeleton (flying dragon) located outside Germany.
- A Squaloraja fish skeleton in 1829.
- Her observations played a key role in the discovery that coprolites, known as bezoar stones at the time, were fossilised faeces.
- She discovered that belemnite fossils contained fossilised ink sacs like those of modern cephalopods.

The late eighteenth century saw Lyme Regis developing into a popular seaside resort with many wealthy and middle-class tourists taking staycations to avoid the French Revolutionary Wars raging on the continent. Fossil collecting became increasingly popular with local people making money by selling 'curios' to visitors in straitened times; the Anning family were avid collectors/sellers, topping up their disposable income, collecting and selling on fossils with catchy names such as 'snake-stones' (ammonites), 'devil's fingers' (belemnites), and 'verteberries' (vertebrae), to which medicinal and mystical properties were attributed. What began first as a pastime for many soon became a science for the educated few as the importance of fossils to geology and biology became increasingly understood. The best time for collecting was in winter when the cliffs could be unstable, with landslides exposing new fossils.

The perils of winter collecting are described well by the *Bristol Mirror*:

> This persevering female has for years gone daily in search of fossil remains of importance at every tide, for many miles under the hanging cliffs at Lyme, whose fallen masses are her immediate object, as they alone contain these

> valuable relics of a former world, which must be snatched at the moment of their fall, at the continual risk of being crushed by the half suspended fragments they leave behind, or be left to be destroyed by the returning tide: – to her exertions we owe nearly all the fine specimens of Ichthyosauri of the great collections…

Anning's mother Molly ran the fossil business after her husband Richard's death, in 1821. By 1825, Mary Anning took control of the family specimen business with much of the trade going through Anning's Fossil Depot with its glass storefront display window attracting numerous visitors, including no less than King Frederick Augustus II of Saxony, who in 1844 purchased an ichthyosaur skeleton for his extensive natural history collection.

Despite not always getting the recognition and credit for her many groundbreaking findings, being shunned by the Geological Society and a lack of funds and funding, Anning was championed by her friend, geologist and palaeontologist Henry de la Beche KCB, FRS (1796–1855), who did her a huge favour when he commissioned Georg Scharf to make a lithographic print based on de la Beche's watercolour painting, *Duria Antiquior*, portraying life in prehistoric Dorset and based largely on fossils Anning had found. De la Beche generously sold copies of the print to his fellow geologists and other wealthy friends, donating the proceeds to Anning. Anning naturally resented the academic plagiarism which she was subjected to.

All the while, Anning's reputation was growing; she was nothing if not studious and dedicated to her work. Palaeontologist Christopher McGowan tells how he examined a copy Anning made of an 1824 paper by William Conybeare on marine reptile fossils and noted that the copy included several pages of her detailed technical illustrations that he was hard-pressed to tell apart from the original.[1] She also dissected modern animals, including both fish and cuttlefish, to gain a better understanding of the anatomy of some of the fossils with which she was working.

This is how Lady Harriet Silvester described Anning in her diary:

> The extraordinary thing in this young woman is that she has made herself so thoroughly acquainted with the science that the moment she finds any bones she knows to what tribe they belong … It is certainly a wonderful instance of divine favour – that this poor, ignorant girl should be so blessed, for by

> reading and application she has arrived to that degree of knowledge as to be in the habit of writing and talking with professors and other clever men on the subject, and they all acknowledge that she understands more of the science than anyone else in this kingdom.[2]

Anna Pinney, an occasional companion on the fossil hunts, wrote about the plagiarism and academic rivalry she suffered: 'She says the world has used her ill … these men of learning have sucked her brains, and made a great deal of publishing works, of which she furnished the contents, while she derived none of the advantages.'[3] Anning herself wrote how the world had used her so unkindly, fearing that it had made her 'suspicious of everyone'.[4]

Anning's experience was, of course, by no means unique: it was commonplace for the finds, endeavours and research carried out by working-class or even just non-academic people in early nineteenth-century scientific literature to be re-assigned to a wealthy collector or paid-up member of an exclusive academic society. Anning, and women like her, were doubly disadvantaged, being working class and not a man. When a fossil had been found by a quarryman, builder or road worker and sold on to a wealthy collector, many of these collectors were unscrupulous enough to claim that they were the finder.

Notes

1. Christopher McGowan, *The Dragon Seekers*, Persius Publishing, 2001, pp. 14–21.
2. 'Mary Anning', University of California Museum of Paleontology.
3. McGowan, 2001, pp. 203–204.
4. Charles Dickens, 'Mary Anning, the Fossil Finder', *All Year Round*, 13, 1865, pp. 60–63.

Ellen Edwards (1810–1889)

Edwards was born on Anglesey and is credited with teaching navigation to more than 1,000 seamen; she picked up her skills, mastering mathematics and navigation under their father's guidance at the school of navigation he, Capt. William Frances Edwards, set up in 1814 at Almwch. By 1830, she had established her own school of navigation in Caernarfon: 'Mrs Edwards, has for a long time kept a school of navigation, and taught the cunning craft to nearly every master mariner of the port.' Ellen Edwards was recognised as 'a clever woman, highly accomplished in the art of navigation, who had taught some of the best seamen to command large vessels'.

Eighteen forty-seven was a bad year for Ellen. Are we that surprised to learn that a report commissioned to assess the state of education in Wales attempted to undermine her achievements? The report, influenced by sectarian rivalries and personal animosity, belittled Ellen's contribution, summing up mean spiritedly: 'All the navigation which has been learned here as a science has been taught by an old woman of Caernarfon.' (She was a spritely 37 at the time!) However, Ellen's supporters, including former students who had achieved great success in their maritime careers, rallied behind her, applauding her immense skill and dedication.[1] The contentious report came to be known as 'Brad y Llyfrau Gleision' or 'Treachery of the Blue Books'.

When she was 70, the Caernarfon Harbour Trust lobbied the Government for her to receive an honorary pension in light of 'outstanding services of this woman'. She received a miserable and miserly £75 from the Royal Bounty Fund for her fifty to sixty years of service. On her death, a newspaper eulogised: 'She was the most successful seamen's teacher in North Wales for the long period of 60 years.'

Notes

1. Michael James, *The Legacy of Ellen Edwards: A Pioneering Maritime Teacher, Caernarfon*, www.visitcaernarfon.co.uk/post/the-legacy-of-ellen-edwards-a-pioneering-maritime-teacher-caernarfon

Augusta Ada King, Countess of Lovelace (1815–1852)

Augusta Ada King, Countess of Lovelace, also known as Ada Lovelace (née Byron) was an English mathematician and writer principally known for her work on Charles Babbage's proposed mechanical general-purpose computer, the Analytical Engine. She was the first to recognise that the machine had applications beyond pure calculation.

King had an eventful early life, being the only legitimate child of Lord Byron and reformer Anne Isabella Milbanke; all her half-siblings, Byron's other children, were born out of wedlock to other women.[1] Byron separated from his wife a month after Ada was born and left England, never to return; he died in Greece when Ada was 8.

Lady Byron, however, was solicitous regarding her daughter's upbringing and encouraged Ada's passion for mathematics and logic so as to prevent her from developing Byron's apparent insanity. Ada, for her part, pursued her studies avidly. She married William King in 1835: King was made Earl of Lovelace in 1838, so Ada thereby became Countess of Lovelace.

Lovelace's socialising and educational endeavours brought her into contact with scientists such as Charles Babbage and Michael Faraday, as well with Charles Dickens. She described her approach as 'poetical science' and herself as an 'Analyst (& Metaphysician)'.[2] By the age of 18 she had begun a long working relationship and friendship with Babbage. She was in particular interested in his work on the Analytical Engine.[3] Lovelace first met him in 1833, when she and her mother attended one of Charles Babbage's Saturday-night soirées with their mutual friend, and Lovelace's tutor, Mary Somerville.[4]

In 1843, Lovelace translated a paper by the military engineer Luigi Menabrea (later Prime Minister of Italy) about the Analytical Engine, supplementing it with notes which described a method of using the machine to calculate Bernoulli numbers which is often called the first published computer program. Fuegi and Francis (2003) tell us that 'She also developed a vision of the capability of computers to go beyond mere calculating or number-crunching, while many others, including Babbage himself, focused only on those capabilities.'[5] Toole (1987) reports that Lovelace was the first to point out the possibility of encoding information besides mere arithmetical figures, such as music, and manipulating it with such a machine. Her mindset of 'poetical science' led her to ask questions about the Analytical Engine (as shown in her notes), examining how individuals and society relate to technology as a collaborative tool.[6] She wrote:

> [The Analytical Engine] might act upon other things besides *number*, were objects found whose mutual fundamental relations could be expressed by those of the abstract science of operations, and which should be also susceptible of adaptations to the action of the operating notation and mechanism of the engine ... Supposing, for instance, that the fundamental relations of pitched sounds in the science of harmony and of musical composition were susceptible of such expression and adaptations, the engine might compose elaborate and scientific pieces of music of any degree of complexity or extent.[7]

This moved thinking about the potential of computing devices to another level and anticipated the implications of modern computing a century before they were realised.

Notes

1. Betty Alexandra Toole, 'Poetical Science', *The Byron Journal*, 15, 1987, 55–65.
2. Betty Alexandra Toole, *Ada, The Enchantress of Numbers: Prophet of the Computer Age*, Strawberry Press, 1998, pp. 156–157.
3. The Analytical Engine was a proposed digital mechanical general-purpose computer designed by Charles Babbage. It was first described in 1837 as the successor to Babbage's Difference Engine, which was a design for a simpler mechanical calculator. The Analytical engine incorporated an arithmetic logic unit, control flow in the form of conditional branching and loops, and integrated memory, making it the first design for a general-purpose computer that could be described in modern terms as Turing-complete.
4. Toole, *Ada, the Enchantress of Numbers*, Location 641.
5. J. Fuegi, and J. Francis, 'Lovelace & Babbage and the creation of the 1843 "notes"' (PDF), *Annals of the History of Computing*, 25(4), 2003, pp. 16–26 (pp. 19, 25).
6. Toole, 'Poetical Science'.
7. Ada Lovelace and Luigi Menabrea (1842), 'Sketch of the Analytical Engine invented by Charles Babbage Esq', *Scientific Memoirs*, Richard Taylor, 694. Rowan Hooper (16 October 2012), 'Ada Lovelace: My brain is more than merely mortal'. *New Scientist.*

Mary Ward (1827–1869)

Mary Ward (née King) was an Irish naturalist, astronomer, microscopist, author, and artist. Sadly, she was killed in 1869 when she was run over by an experimental steam car built by her cousins. This makes her the first person known to have been killed by a motor vehicle.

Home taught by her scientific parents, she was interested in nature from a very young age and was avidly collecting insects by age 3. Her father bought her a compound microscope to help her with her detailed drawings of insects; the Andrew Ross (model 112) cost him £48 12s 8d but it was money well spent as it instilled in her a passion for all things microscopical and she read everything relating to microscopy until she was expert in the field. Excluded from universities and learned societies because she was a woman, although an expert woman, her cousin William Parsons was made president of the Royal Society, which enabled her to meet countless scientists.

Ward was also an enthusiastic astronomer. She was one of only three women on the books of the Royal Astronomical Society; the others were Queen Victoria and Mary Somerville, after whom Somerville College at Oxford University was named. Her cousin, William Parsons, 3rd Earl of Rosse, built the world's largest reflective telescope at Birr Castle (where it remained the largest in the world for decades).

Ward's first book, *Sketches with the Microscope*, was privately printed in 250 copies in 1857; she believed that no academic publisher would touch it because of her gender and lack of academic credentials. She also ran off several hundred flyers – an excellent example of early book marketing. The 250-copy printing sold out in weeks – enough to make a London publisher take the risk and contract for future publication. The book was reprinted eight times between 1858 and 1880 as *A World of Wonders Revealed by the Microscope*. Her other books are: *A Windfall for the Microscope* (1856), *A World of Wonders, Revealed by the Microscope* (1857), *Entomology in Sport, and Entomology in Earnest* (1857, with Lady Jane Mahon), *Microscope Teachings* (1864), and *Telescope Teachings* (1859). She illustrated her books and articles herself, as well as many books and papers by other scientists.

Mary Pirie (1822–1885)

Mary Pirie was a fervent botanist, entomologist and teacher born in Aberdeen and one of ten children. She had a lifelong passionate interest in the benefits of education and science for all, describing entomology as the study of 'animated nature' – the key focus of her second book, *Familiar Teachings on Natural History: A Book for the Use of Schools and Families* (1864). Four years earlier, in her first book, she had asserted that one of the great benefits of natural sciences was the way that it is accessible to the masses, stating: 'The Science of Botany is the branch of Natural History most generally accessible to all; it is the science which furnishes us with the key to unlock the cabinet and unfold the wonders of the Floral Kingdom.'[1] Accordingly, she published about botany in an accessible and digestible way through poetry and illustrations.

The Biographical Dictionary of Scottish Women tells us that 'Pirie was a botanist who wanted to inspire that love in others, both her works are aimed at encouraging people from all walks of life to be interested in nature.'[2]

Pirie's 1860 popular work is acknowledged as 'a beautiful book for a present'.[3] It includes her own poetry and illustrations. One shining example of Pirie's accessible style is in the details of the romantic story she tells about the mignonette, reseda odorata, and how it became part of the Count Walsheim of Saxony's coat of arms.

Notes

1. M. Pirie, *Flowers, Grasses, and Shrubs: A Popular Book on Botany, Illustrated*, pp. 157–161.
2. Elizabeth Ewan et al., *The Biographical Dictionary of Scottish Women: From the Earliest Times to 2004*, Edinburgh, 2006, pp. 292.
3. 'A beautiful book for a present', *Bell's Weekly Messenger*, 20 October 1860.

Dame Maria Gordon (1864–1939)

Maria Gordon was born in Monymusk, Aberdeenshire. After education in Edinburgh and London, she continued her geological studies at the University of Munich, specialising in recent and fossil corals.

In 1900, she became the first woman to gain a PhD from there. She received a distinction in geology, palaeontology and zoology.

In 1895, she married John Gordon, an Aberdeen physician and loyal supporter of her career. She then went on to have three children. 'Even as a busy mother, she continued her fieldwork, often trekking alone in the mountains, making observations and collecting fossil specimens.'

Geological work

All Maria Gordon's studies were carried out in the geologically complex south Tyrol Dolomites – a mountain range in the northern Italian Alps and now in the UNESCO World Heritage List. It was at that time a remote and isolated region.

Her early work was notable particularly for her theory of 'crust-torsion'. This revealed that the distinctive peaks of the Dolomites were formed by pushing, folding and twisting movements of the Earth's crust. Her ideas contradicted the prevailing view of the day – that the peaks were formerly coral atolls in an ancient sea.

Over the years, she published more than thirty original papers on the geology of this region. Some of these became definitive references for future work in the south Tyrol.

Her writing brought her wide professional recognition, first in Austria, then Italy. The London Geological Society awarded her the Lyell medal in 1932.

Women's rights campaigner

Throughout much of her life, Maria Gordon was also an active campaigner for women's rights, encouraging women to get politically involved.

After the death of her husband in 1919, she moved to London and became active in civic affairs. She was made a Justice of the Peace and was the first woman to chair a London borough court.

In 1935, she was made a Dame, which recognised her tireless campaigning for the rights of women and children. It also recognised her involvement in women's groups such as the International Council of Women which is still active today.

Jessie Chrystal Macmillan (1872–1937)

Chrystal Macmillan was a woman of many achievements: suffragist, peace activist, barrister, feminist and the first female science student to graduate from the University of Edinburgh and the university's first female honours graduate in mathematics. She was the second woman to plead a case before the House of Lords, and was one of the founders of the Women's International League for Peace and Freedom.

Eighteen ninety-two saw Macmillan among the first female students to enrol at the University of St Andrew's, where she studied science subjects including Honours Mathematics with George Chrystal (no relation), Astronomy with Ralph Copeland, and Natural Philosophy with Peter Guthrie Tait and Cargill Gilston Knott. In April 1896, as just noted, she graduated with a BSc with first-class honours in mathematics and natural philosophy, the first woman at the university to do so.[1]

In 1896, she went to Berlin to further her university studies and on returning to Edinburgh passed an examination in Greek language to enter the Faculty of Arts in October 1896. She then studied a number of subjects including politics, graduating in April 1900, also earning second-class honours in Moral Philosophy and Logic. During this time she was a member of the Edinburgh Ladies' Debating Society; she also joined the Edinburgh Mathematical Society in May 1897, the second woman member after Flora Philip in 1896.

Notes

1. J. J. O'Connor and E. F. Robertson (January 2008), 'Jessie Chrystal Macmillan'. *MacTutor History of Mathematics*. University of St Andrews Scotland, School of Mathematics and Statistics.

Medical Doctors

The *Westminster Review* of 1868: 'Men have never made an outcry against women's entering upon any occupation however hard or "degrading," unless that occupation were one in which they would compete with men!'[1]

The Universities of Oxford (ca. 1167) and Cambridge (1209) were the only two English universities until University College, London (1827) and King's College, London (1829) combined to form the University of London in 1836; the University of Durham was founded in 1832. They all had one thing in common: none accepted women as students, thus barring the brightest of half the country's population from trying to better themselves and from improving the lives of the ill and infirm. They could, however, study for degrees and pass the requisite examinations but they were not actually awarded the degrees they had been working towards.

The women in this important section of the book were all greatly disadvantaged by the fact that the innately insecure sexist male bastion that characterised the medical profession flatly denied access to women.

In 1867, a pioneering and inspirational Anne Clough changed all that when she established the North of England Council for Promoting Higher Education for Women (NECPHEW) which organised university lectures for women. Jane Jordan tells how at its first meeting, Ladies' Educational Associations were represented by Elizabeth Wolstenholme from Manchester and Lucy Wilson from Leeds. Clough became honorary secretary and Josephine Butler took office as its president, a position she held until she stood down in 1871. Wolstenholme drew up the rules and her friend, the academic James Stuart, gave a series of lectures for women in astronomy. F. W. H. Myers was another tutor; more than 550 students signed up to his university lectures.[1]

Subsequently, degrees were awarded to women from 1878 at the University of London, 1895 at Durham, 1920 at Oxford, and 1948 at late-to-the-party Cambridge.

In 1929 Virginia Woolf (1882–1941), looking back, remarked on the wealth built up by the men's ancient colleges, evident from their impressive buildings, none of which ever impacted the development of women's education.

Emily Davies (1830–1921) knew all about the problems of female accessibility to universities when fundraising in the 1870s to build a women's university college offering the same examinations as male undergraduates at Cambridge.

> We are told that we ought to ask for £30,000 at least ... It is not a large sum, considering that there is to be but one college of this sort ... and considering how easy it is to raise immense sums for boys' schools. But considering how few people really wish women to be educated, it is a good deal.

Emily Davies (1830–1921)

In October 1868, a University Diploma secured the professional accreditation of women educators.

By 1872, Manchester Ladies' Educational Association was represented by Elizabeth Gaskell's daughter Meta Gaskell and the Ladies' Educational Association in Leeds by Frances Lupton who was supported by her sister-in-law, Anna Lupton, and Lord Houghton. Some of the council's members also belonged to the Education for Girls Committee of the Royal Society of Arts which, from 1871, had aligned itself with the aims of the NECPHEW.

Notes
1. Jane Jordan, *Josephine Butler*, pp. 104, 87–88.

Elizabeth Garrett Anderson (1836–1917)

Elizabeth Garrett Anderson was the first woman in Britain to qualify as a physician and surgeon in 1865 after exposing a loophole in the Worshipful Society of Apothecaries' admissions process, passing her exams first time. But that was only the first barricade she had to negotiate and clamber over: not one British hospital would admit her as a doctor and so, what did she do about it? She set up her own practice on Euston Road, run by women, for women.

> She became an inspiration for women everywhere, proving their worth in the medical profession. Her life was one of many pioneering firsts, including becoming the first female Mayor, the first female Dean of a medical school, and the first woman to ever sit on a school board of governors in Britain. Her example went a long way to liberalising admission policy at medical schools across the country: in 1876, partly due to Elizabeth's open campaigning, an act was passed forcing the British Medical Register to accept women. By 1914, there were over 1000 female doctors in England. However, not everyone was happy … about this, and the struggle for equal standing in the profession continued.[4]

Garrett Anderson was not the only household name who passed the Apothecaries' exam. Agatha Christie passed the Apothecaries Hall Assistants' Exam in 1917, an achievement which served her well in her subsequent literary career as a crime writer, furnishing as it did a familiarity with poisons and their modes of actions, antidotes and so on. This exam had become popular with many women who wanted to work as dispensers (pharmacy technicians today) but who did not want to sit the Pharmaceutical Society exams. Elizabeth Garrett Anderson took on pupils to train under her dispenser at her St Mary's Dispensary for Women and Children, first opened in 1866.

By the late nineteenth century, educated women generally began to attend colleges, including medical schools, and to access the professions. The first English woman doctor was Elizabeth Blackwell, who obtained her degree and practised medicine in the United States where the 'medical women movement … was at least 20 years ahead of its British counterpart'.

Increasingly, women were working as physicians or other medical professionals but were denied the associated official licences and degrees. The Medical Act 1858 was an attempt to eliminate this practice by forcing licensed doctors to register with the government. Women sidestepped this by studying medicine in Europe as a way to becoming licensed doctors in Britain. It would, however, be seven more years before the next woman was granted a medical licence; it was granted to Sophia Jex-Blake and three others belonging to the Edinburgh Seven, the first seven women allowed entry into the University of Edinburgh's medical school; of the Edinburgh

Seven, these first four who registered as doctors had actually graduated from the Irish College of Physicians in 1877.

Another stumbling block to entry into the profession was the vexed question whether it was, in the first place, appropriate for women to obtain medical degrees and then go on to practise, which obviously involved tending to women *and* men. The debate took place not only in Parliament and in universities but also in the home where young women had to convince their mothers and fathers to back their career choice and the attendant apparent moral dangers; it took place in opinion-forming medical journals like *The Lancet* and the *British Medical Journal*, and it took place in popular papers and magazines such as *The Times* and *Punch*.

Location of halls of residence became an important issue and had to be addressed effectively and sensitively – would our girls be safe from predatory men? For example, 'respectability' at Cambridge was ensured by choosing Benslow House in the (then remote) village of Hitchin, between Cambridge and London, to deter ill-intentioned visits by male students from Cambridge. In London itself, the British Federation of University Women purchased a lease for Crosby Hall, in fashionable Cheyne Walk, London, SW3 in 1926 as an international hall of residence for university women, with a purpose-built extension.

All things considered, by being a doctor in the Industrial Revolution, all a woman was doing was what she had been doing for millennia already – administering to the health needs of the family; women had an historical role as healers in all cultures and societies and were good at healing – the only differences being that now she would be paid for it and she would be administering outside the home to strange men, not just husband, brother or son. The prospect of this alarmed many a traditionalist and, added to the fact that these newly independent women would be working outside the home and putting patients before family, it is no wonder that anxieties regarding the role of women in society and other established values, not least femininity and masculinity, flourished unabated at the end of the nineteenth century. As quoted at the beginning of the chapter, there is also the not insignificant fact that many men may have felt threatened professionally by these cohorts of newly trained, enthusiastic and aspirational women.

At the same time, women were not just eager to learn how to apply medical science in a diagnostic and therapeutic setting; they, as a sex, had an existential stake in advancing medical knowledge in order to advance women's health generally in order to help other women who had received, and to our shame, still get in certain areas of medicine, short shrift when it came to medical research and training, particularly in obstetrics and gynaecology, the development of surgical techniques, cutting-edge diagnoses and therapeutics.

The advancement of women's health as a specialty in its own right surely seemed an unobtainable and remote prospect if left in the hands of the eighteenth and nineteenth-century male, sometimes sexist doctor. The medical issues from which British women, all women, suffered – for example, STIs like gonorrhoea and syphilis (often spread by men to their wives or partners via infectious prostitutes), high neonatal mortality and what today we term obstetrics and gynaecology were largely ignored and underfunded, not to say masculinised by replacing experienced midwives with 'man-midwives'.

Being doctors with a determination to advance women's health, women, in theory at least slowly gained access to and experience in history taking, patient examination, signs and symptoms, therapeutics, medical technology, patients, visitors, hospital scenarios, education, and the support of some enlightened men in the profession – none of which would have been a stranger to male doctors.

In a nation which fostered and depended on its empire, it is inevitable that our medical protocols and systems impacted many women within that vast empire in different cultures and religions. Queen Victoria had observed that many of her female Indian subjects had medical issues for which they could not be treated because of purdah, the Hindu practice of secluding women, which often deprived them of medical care when the only doctors available were men. To illustrate the crisis that this was, we note that in their bibliography of women in medicine, Chaff, Haimbach, Fenichel, and Woodside reference fifty-five articles and books on medical missionary work written *by* women physicians and published during the last decade of the nineteenth century.[5] For more on this, see Francina Ford-Sorabji and Dame Mary Scharlieb.

Notes

1. *The Westminster Review*, 34 (1868), p. 458.
2. This follows publication of the GMC's *The state of medical education and practice: Workforce report* in November 2024.
3. The GMC's 2024 *National training survey* measured the experience of doctors in training in the UK. Of those who answered optional questions on discriminatory behaviours in the workplace, nearly one in ten women (9 per cent) reported experiencing unwelcome sexual comments, or advances causing embarrassment, distress or offence, compared to just 4 per cent of men. The proportion rose drastically for women at the beginning of their training. Nearly a fifth (18% per cent) of women in their first year of foundation training reported such experiences, 13 per cent in their second year and 6 per cent at higher training levels.
4. www.cwplus.org.uk/women-in-medicine
5. Sandra L. Chaff et al., *Women in Medicine: A Bibliography of the Literature on Women Physicians*, Metuchen, N. J.

Joanna Stephens (1692–1774)

Eighteenth-century quack or medical pioneer?

In 1739, Joanna Stephens was paid £5,000 by the British Government for a recipe to cure bladder stones; an astonishing amount of money at that time, equivalent to around £650,000 today.

The British Association of Urological Surgeons tells us that

> The historical treatment of bladder stone was open lithotomy, an agonising, dangerous operation; thus, any non-surgical alternative to treat stones was eagerly sought. So, when in 1738 an article was published in the *Gentleman's Magazine* extolling the benefits of Mrs Stephen's Cure for the stone, national interest was generated.[1]

Stephens offered to sell her recipe to the country for £5,000. At first, a private subscription was set up through the *Gentleman's Magazine* and, when this only raised £1,356 3s, a petition was made to Parliament. A Parliamentary committee was set up and they tested the recipe on four stone sufferers. The patients were examined before and after by a committee of twenty-two doctors, politicians and scientists and found to be cured. Parliament agreed to pay Joanna Stephens £5,000 and she published her recipe in the *London Gazette* on 16 June 1739.

The powder potion consisted of eggshells and snails; the decoction of boiled herbs, soap, swine's cresses, and honey; and the pills, snails, wild carrot seeds, burdock seeds, ashen keys, hips, haws, soap and honey.

As a young woman, Stephens reputedly concocted remedies that she distributed free of charge to poor people. After the death of a close friend from kidney stones, she devoted herself especially to remedies against stones. Having discovered, by chance, a recipe for the dissolution of stone in the bladder and kidneys, she began to prepare and dispense it. The evidence for this is not terribly empirical, though.

Accordingly Joanna Stephens has, by many, been demoted to the status of 'quack'. The bad news is that she *does* feature in Paul Strathern's *Quacks, Rogues and Charlatans* published by the Royal College of Physicians (p. 59). However, the British Association of Urological Surgeons (BAUS) concludes it is possible

> that her remedy would have worked for some patients and, more importantly, it stimulated the pursuit of chemical analysis of stones which subsequently formed the basis of modern medical biochemistry. It was also said that she 'disappeared' once she had been paid by the government, again suggesting she was a trickster. But it appears she merely bought a nice house in Brookgreen, in Hammersmith where she lived to the age of 82.

Notes

1. www.baus.org.uk/museum/1438/joanna_stephens

Margaret Anne Bulkley; aka James Barry (c. 1789–1865)

James Barry was a trailblazing and eminently successful man in many ways. James Barry, however, was born a woman and so qualifies as 'a trailblazing, and eminently successful *woman* in many ways'.

Writing in the *New Scientist*, Stephanie Pain reveals that

> Barry obtained a medical degree from the University of Edinburgh Medical School, then served first in Cape Town, South Africa, and subsequently in many parts of the British Empire. Before retirement, Barry had risen to the rank of Inspector General (equivalent to Brigadier) in charge of military hospitals, the second-highest medical office in the British Army. He improved conditions

> not only for wounded soldiers, but also those of the native inhabitants. Barry performed the first recorded caesarean section by a European in Africa in which both the mother and child survived the surgery.[1]

We learn from the BBC that

> Barry, who lived his entire adult life as a man, was named Margaret Anne at birth and was known as female in childhood. Barry lived as a man in both public and private life, at least in part, in order to be accepted as a university student, and to pursue a career as a surgeon. His anatomy became known to the public and to military colleagues only after his death.[2]

Unsurprisingly, the non-military aspects of Barry's life are embellished with speculation. However, we do know that a plan was hatched to enable Barry to enter the University of Edinburgh. Margaret Anne Bulkley then became James Barry, nephew of the late James Barry RA, and remained known as such for the next fifty-six years. Du Preez (2016) tells that

> His short stature, slight build, unbroken voice, delicate features and smooth skin led others to suspect that he was not a man but a pre-pubescent boy. This identity was maintained through surgical training and recruitment into the British Army which, at officer rank level, did not then require a medical examination.

During the Crimean War (1854–1856), Barry got embroiled in an argument with Florence Nightingale. After his death, Nightingale wrote peevishly that:

> I never had such a blackguard rating in all my life – I who have had more than any woman – from this Barry sitting on [her] horse, while I was crossing the Hospital Square, with only my cap on, in the sun. [He] kept me standing in the midst of quite a crowd of soldiers, commissariat servants, camp followers etc etc, every one of whom behaved like a gentleman, during the scolding I received, while [she] behaved like a brute. After [she] was dead, I was told [he] was a woman. … I should say [she] was the most hardened creature I ever met.

Barry would never allow anyone into the room while undressing, and repeated a standing instruction that 'in the event of his death, strict precautions should be adopted to prevent any examination of his person' and that the body should be 'buried in [the] bed sheets without further inspection'. We must agree that with Kubba and Young that 'Concealment of one's sex for ... 40 years in the British Army, is simply unbelievable'. Theories that Margaret Anne Barry cannot have been a woman because someone born female could not have achieved as much as Barry did are disappointing, if predictable, with one biographer writing, 'Dr. Barry couldn't have been a woman, for women and medicine were contradictory terms ... it was still too much to imagine that any female could perform as brilliantly as Dr. Barry had done.'

Notes

1. Stephanie Pain (6 March 2008), 'The "male" military surgeon who wasn't', NewScientist.com.
2. 'Five British heroes overlooked by history', *BBC News*, 17 November 2009.

Elizabeth Blackwell (1821–1910)

Elizabeth Blackwell is famous for being the first woman to earn a medical degree in the United States, and the first woman on the Medical Register of the General Medical Council for the United Kingdom. She played a crucial role in both the United States and the United Kingdom as a social reformer, and was a pioneer in promoting education for women in medicine. Her contributions are celebrated by the Elizabeth Blackwell Medal, awarded annually to a woman who has made a significant contribution to the promotion of women in medicine.

Up to around 1868, Blackwell was in the US, learning and teaching medicine, forever emphasising the importance of educating young women, notably medical students. She published *The Laws of Life with Special Reference to the Physical Education of Girls*, her first work, a book about the physical and mental development of girls, focusing on the preparation of young women for motherhood.

Sexism and male obstructionism seemed to stalk her wherever she went, in the USA and in Europe. Indeed, most physicians advised that she either go to Paris to study or else disguise herself as a man to study medicine. The main reasons for her rejection were that (1) she was a woman and therefore intellectually inferior, and (2) she might actually prove equal to the task, prove to be competition, and that she could not expect them to 'furnish [her] with a stick to break our heads with'.

In April 1849, Blackwell decided to continue her studies in Europe, visiting hospitals in Paris. In June, Blackwell enrolled at La Maternité, a 'lying-in' hospital, on condition that she would be treated as a student midwife, not a physician. In 1857, she opened the New York Infirmary for Indigent Women and Children with her sister Emily, who had also become a doctor. The Infirmary was staffed by women, with a mission to employ female physicians. During the American Civil War, they trained nurses.

Due to a clause in the Medical Act 1858 that recognised doctors with foreign degrees practising in Britain before 1858, she became the first woman to have her name entered on the General Medical Council's medical register on 1 January 1859.[1] She also became a mentor to Elizabeth Garrett Anderson during this time. By 1866, nearly 7,000 patients were being treated every year at the New York Infirmary, and Blackwell was needed back in the United States.

In 1874, Blackwell set up a women's medical school in London with Sophia Jex-Blake, who had been a student at the New York Infirmary some years earlier. Blackwell thought Jex-Blake was dangerous, belligerent, and tactless.[2] Nonetheless, Blackwell became deeply involved with the school, which opened in 1874 as the London School of Medicine for Women, with the main aim of preparing women for the licensing exam of Apothecaries Hall. Blackwell vehemently opposed the use of vivisections in the laboratory of the school.[3]

After the establishment of the school, Blackwell lost much of her authority to Jex-Blake and was elected as a lecturer in midwifery. She resigned this position in 1877, officially retiring from her medical career.

While Blackwell viewed medicine as a means for social and moral reform, her student Mary Putnam Jacobi focused on curing disease. At a deeper level of disagreement, Blackwell felt that women would succeed in medicine because of their humane female values, but Jacobi believed that women should participate as the equals of men in all medical specialties.

Blackwell moved to Britain in 1869 where she focused on social reform and writing. She co-founded the National Health Society in 1871 which aimed to educate people about the benefits of hygiene and healthy lifestyles. Her friend Barbara Bodichon introduced Blackwell into her circles.

Social reform

Blackwell invested in the founding of two utopian communities: Starnthwaite and Hadleigh, Essex in the 1880s. She believed that Christian morality ought to have as large a role as scientific inquiry in medicine and that medical schools ought to instruct students in the subject. She distrusted the value of inoculation and thought it dangerous. She believed that bacteria were not the only important cause of disease and felt their importance was being exaggerated.

Blackwell campaigned against licentiousness, prostitution and contraceptives, arguing instead for the rhythm method of birth control. She campaigned against the Contagious Diseases Acts, arguing that it was a covert legalisation of prostitution. Her 1878 book *Counsel to Parents on the Moral Education of their Children* argued against the Act. Controversially, she believed women's sexual libidos to be equal to those of men, and that men and women were equally responsible for controlling their sexual urges.

Notes

1. Enid Moberly Bell, *Storming the Citadel: The Rise of the Woman Doctor*, London, pp. 25, 1953.
2. Ibid.
3. Nancy Ann Sahli, *Elizabeth Blackwell, M.D. (1871–1910): A Biography*, New York: Arno Press, 1982.

Isabel Thorne (1834–1910)

Isabel Jane Thorne (née Pryer) was an early campaigner for medical education for women. Mrs Thorne was a member of the feminist Edinburgh Seven, who campaigned and succeeded in securing the right in law for women to qualify as doctors. While living in Shanghai with her tea merchant husband, one of her five children died, inspiring her to travel in China and India to help women and children convinced of the need for women to have female doctors for themselves and their children.

Thorne travelled through China during the Taiping Rebellion.[1] On returning to London in 1868, she started midwifery training at the Ladies' Medical College in Fitzroy Square, later describing the teaching there as 'inadequate'.[2]

She responded to Sophia Jex-Blake's advertisement calling for women to join her in an attempt to qualify as doctors at the University of Edinburgh and so Thorne became one of the Edinburgh Seven. About this time, she won first prize in an anatomy examination. After their attempt to graduate in medicine was blocked, Thorne was one of the women who re-grouped at the London School of Medicine for Women. Her diplomatic temperament meant she was a more acceptable honorary secretary than Jex-Blake, whose nomination was potentially controversial.

Thorne renounced her own ambition to be a doctor in order to commit herself to helping the school run smoothly; to become more solidly established. She was regarded as considerate, tactful, hard-working, industrious, and for thirty years the mainstay of the college administration.[3]

Mrs Thorne kept records and wrote an account of these years which was published in 1905 as *Sketch of the Foundation and Development of the London School of Medicine for Women*. In 1908, her daughter, the surgeon May Thorne, who had graduated from the LSMW in 1895, succeeded her as honorary secretary.

Notes

1. The Taiping Rebellion was a civil war in China between the Qing dynasty and the Taiping Heavenly Kingdom. The conflict lasted fourteen years, from its outbreak in 1850 until the fall of Taiping-controlled Nanjing in 1864. The last rebel forces were defeated in August 1871. Estimates of the conflict's death toll range between 20 million and 30 million people, representing 5–10 per cent of China's population at that time.
2. Veronique Molinari, '"Schools of their own" – The Ladies' Medical College and London School of Medicine for Women'. In Andreolle and Molinari (eds.), *Women and Science, 17*th *Century to Present: Pioneers, Activists and Protagonists*. Part II, Chapter 2. Cambridge Scholars Publishing, pp. 99–124.
3. Shirley Roberts, *Sophia Jex-Blake: A Woman Pioneer in Nineteenth-Century Medical Reform*, London, pp. 144–145.

Sophia Jex-Blake (1840–1912)

A leading light in the seemingly endless struggle to win women the right to a viable university education, when she and six other women – the Edinburgh Seven – embarked on a medical course at the University of Edinburgh in 1869. She wrote to her friend Lucy Sewall, 'It is a grand thing to enter the very first British University ever opened to women, isn't it?'

She was not only the first practising female doctor in Scotland, and one of the first in the United Kingdom, but was also involved in founding two medical schools for women, in London and Edinburgh, at a time when no other medical schools saw fit to train women.

Home educated until age 8, she then attended various private schools in the south of England, and in 1858, enrolled at Queen's College, London, despite her parents' opposition. In 1859, while still a student, she was offered a post as mathematics tutor at the college where she stayed until 1861, living for some of that time with Octavia Hill's family.[1] She worked as a volunteer; in tune with the middle-class values of the time her family did not expect their daughter to earn a living, and her controlling father refused her permission to accept a salary.[2]

In 1865, at the end of the Civil War, Jex-Blake travelled to the United States to learn more about women's education. At the New England Hospital for Women and Children in Boston, she met one of the country's pioneer female physicians, Dr Lucy Ellen Sewall, who became an important and lifelong friend; Jex-Blake worked there as an assistant, during which time she knew she had to be a doctor.

In 1869 she delivered her landmark essay 'Medicine as a profession for women' which was published in a book edited by Josephine Butler: *Women's Work and Women's Culture*.[3] In this, she argued that natural instinct leads women to concern themselves with the care of the sick. With the current situation where education of girls was restricted to domestic crafts, though, women generally had little chance of competing with men as doctors. She further argued that no objective proof existed of women's intellectual inferiority to men. She concluded that the matter could easily be tested by granting women 'a fair field and no favour' – teaching them just as men were taught and subjecting them to the same examinations.

But the joy of enrolling on the medical course was short lived.

The Surgeon's Hall riot

The *Oxford Dictionary of National Biography* tells how when the women demonstrated that they could compete on equal terms with the male students, hostility towards them grew and grew: obscene letters were sent; they were stalked home, had fireworks attached to their front doors, and had mud thrown at them. It all came to an ugly head in the Surgeons' Hall riot on 18 November 1870, when the women arrived to sit an anatomy examination at Surgeons' Hall, and a very angry mob of over 200 gathered outside throwing mud, rubbish, and insults at the women. It is staggering to think that many of this mob comprised supposedly enlightened and intelligent people; so pronounced was the sexism, misogyny even, driving them on in a supposedly enlightened century.

Even though these disturbing events made national headlines and won for the women many new supporters, influential and self-interested members of the medical faculty eventually persuaded a weak and hypocritical university to

> refuse graduation to the women by appealing decisions to higher courts. The courts eventually ruled that the women who had been awarded degrees should never have been allowed to enter the course. Their degrees were withdrawn and the campaign in Edinburgh failed in 1873. Many of the women went to European universities that were already allowing women to graduate, and completed their studies there.[4]

What did Sophia Jex-Blake do next? She certainly didn't give up in the fight for women's medical education.

The British Medical Journal, amongst others,[5] reports that the Medical Act 1876 soon followed, which was an Act to repeal the previous statute, while also permitting medical authorities to license all qualified applicants whatever their gender. The first organisation to take advantage of this new legislation was the Royal College of Physicians of Ireland, but before Jex-Blake applied to them, she passed the medical exams at the University of Berne, where she was awarded a medical doctorate in January 1877. Four months later, she had further success in Dublin and qualified as licentiate of the King and Queen's College of Physicians of Ireland, meaning she could at last be registered with the General Medical Council, the third registered woman doctor in the UK.

On her return to Edinburgh, Jex-Blake opened the Edinburgh Hospital and Dispensary for Women. This was Scotland's first hospital for women staffed entirely by women and offering medical care to poor women for a few pennies. In 1886, she established the Edinburgh School of Medicine for Women,[6] staffed largely by 'a small group of pro female male physicians linked to the University of Edinburgh giving extramural classes open to men and women (which the university could do nothing about). The first students included Elsie Inglis, Grace Ross Cadell, and her sister Georgina, but Jex-Blake's skill as a teacher did not match her role as a doctor.'

A bitter split ensued with her students, culminating in an infamous court case in 1889, where Jex-Blake was successfully sued for damages. The *New Biographical Dictionary of Scottish Women* tells us that

> Thereafter, the Cadell sisters pursued their studies with the more genial, though far younger, Elsie Inglis, who had set up a rival school, the Edinburgh College of Medicine for Women. Jex-Blake's school came to an effective end in 1892, when the University of Edinburgh began taking female students. The Elsie Inglis College continued until 1916, when it merged with the School of Medicine of the Royal Colleges of Edinburgh at Surgeons' Hall.

As a welcome footnote, the Edinburgh Seven were awarded the posthumous honorary MBChB degrees at the University of Edinburgh's McEwan Hall on 6 July 2019. The degrees were collected on their behalf by a group of current students at Edinburgh Medical School.

Notes

1. A private day school for girls aged 11–18 in the City of Westminster, founded in 1848 by theologian and social reformer Frederick Denison Maurice In 1853, it was the first girls' school to be granted a Royal Charter for developing women's education.
2. Margaret Todd, *The Life of Sophia Jex-Blake*, Macmillan, 1918. According to Virginia Woolf, this was a 'typical instance of the great Victorian fight ... of the daughters against the fathers', where a father would hope to keep a daughter in his power by saying earning a living was 'beneath her'. See Chapter 3 of *Three Guineas*.
3. Josephine Elizabeth Grey Butler, *Woman's Work and Woman's Culture: A Series of Essays*, London, 1869.
4. Shirley Roberts, 'Blake, Sophia Louisa Jex- (1840–1912)', *Oxford Dictionary of National Biography*, Oxford University Press, 2004.
5. For example: Great Britain. Parliament. House of Commons (1892). Parliamentary Papers, House of Commons and Command. H.M. Stationery Office. pp. 40ff. John A. Wagner Ph.D. (25 February 2014). *Voices of Victorian England: Contemporary Accounts of Daily Life*, ABC-CLIO, pp. 211f.
6. Edythe Lutzker, *Women Gain a Place in Medicine*, New York: McGraw Hill, p. 149, 1969.

Emily Bovell (1841–1885)

She is recorded as one of the original members of the Edinburgh Seven who worked at the New Hospital for Women in Marylebone Road, London and in Paris, qualifying as a doctor in 1877. The subject of her medical thesis was 'Congestive Phenomena following Epileptic and Hystero-epilectic Fits'. The French awarded her the Officier des Ordre des Palmes Académiques for services to medicine.

Bovell married the physician William Allen Sturge in Paris in 1877; they returned to London together and set up a practice together in Wimpole Street. Bovell went back to Queen's College, lecturing on physiology and hygiene, and running ambulance classes for women.

The *BMJ* obituary for Sturge tells us that her husband was a strong supporter of her professional career, and the cause of women's medical education in general.[1] Five of his sisters had campaigned for women's higher education, including Emily Sturge.

In 1881, her deteriorating poor health caused the couple to give up their practice in London, and move to Nice where she was the first woman doctor; she also campaigned to improve public health and to improve the sewers there.

Notes

1. T. Barlow (12 April 1919), 'Obituary – William Allen Sturge', *British Medical Journal*, 1 (3041), 468–469.

Louisa Atkins (1842–1924)

One of the first British women to qualify in medicine and also England's first female House Surgeon, at the Birmingham and Midlands Hospital for Women. The *BMJ* described her as 'a pioneer in the cause of medicine as a profession for women'.[1]

Atkins had a 'continental upbringing' – she had been educated in France. When still young, she went to India, where in 1860 she married Frederick William Maclean Atkins – an army officer much older than her who died a few years after their marriage when Louisa was still considered a minor. She attended medical school in Zurich, matriculating in 1867 and gaining her degree in 1872. She was one of the 'Zurich Seven': the first group of women to be admitted to, and awarded degrees from, Zurich's medical school.[2]

Her thesis was titled *Ueber Gangraena Pulmonum bei Kindern* (Pulmonary Gangrene in Children).

Catriona Blake, in her *Charge of the Parasols*, describes how women were not yet allowed to sit the examinations of any of the medical licensing bodies, Atkins was not fully registered, but the hospital's committee awarded her the position over two male candidates regardless. Unsurprisingly, this caused some controversy which gave rise to accusations of a fixed appointment.[3] A year later, though, it was acknowledged as a great success, and, Claire Brock said, 'A fair field and no favour should be given to a competent lady candidate.'[4] In 1877, Atkins took the Licence of the King and Queen's College of Physicians in Ireland when they started admitting women, qualifying alongside Sophia Jex-Blake and Edith Pechey.

Atkins was an early member of the Association of Registered Medical Women (ARMW), formed in 1879 in response to the British Medical Association's refusal to admit women doctors. At that time, fourteen women were on the medical register, all of whom were invited to join the Association. Fellow members included Elizabeth Blackwell, Elizabeth Garrett Anderson, Sophia Jex-Blake, Annie Reay Barker, Annie Clark, Matilda Chaplin Ayrton, Eliza Macdonogh Frikart, and Eliza Walker Dunbar.

Atkins began work at the New Hospital for Women (later the Elizabeth Garrett Anderson and Obstetric Hospital), where her colleagues included Elizabeth Garrett Anderson and Annie de la Cherois. She resigned from the New Hospital in 1888, due to her concerns over Garrett Anderson's surgical proficiency.[5] Edith Pechey was appointed her successor.

Notes

1. British Medical Journal Publishing (11 January 1924), 'Louisa Atkins, M.D', *British Medical Journal*, 2 (3331), 836–837.
2. The first seven women who were awarded medical degrees at the university were known as the Zurich Seven. Including Suslova, they were: Nadezhda Suslova (Russia); Frances Elizabeth Morgan (England); Louisa Atkins (England); Maria Bokova (Russia); Eliza Walker (Scotland); Susan Dimock (United States); Marie Vögtlin (Switzerland). Thomas Neville Bonner, *To the Ends of the Earth: Women's Search for Education in Medicine*, Harvard University Press, pp. 32–41, 1992.
3. Lucy Bland, *Banishing the Beast: Sexuality and the Early Feminists*, New York, 1995.
4. Claire Brock, *British Women Surgeons and Their Patients, 1860–1918*, Cambridge, 2017.
5. J. F. Geddes, '"Too high a percentage of failures"? Cover-up at the New Hospital for Women', *Journal of the Royal Society of Medicine*, 103(9), 348–351, 2010.

Frances Hoggan (1843–1927)

Frances Elizabeth Hoggan (née Morgan) in 1870 became the first woman from the UK to receive a doctorate in medicine from any university in Europe. She is remembered as a pioneering medical practitioner, researcher and social reformer – and the first female doctor to be registered in Wales. She opened the first husband-and-wife medical practice in Britain.

As a teenager, she gave birth to an illegitimate daughter, who was brought up with her mother and passed off as Frances's sister. She went on to study at Paris and Düsseldorf.

Excluded from university medical courses, the only route open for women to obtain a licence to practise medicine in England and Wales was, as we know, through the examining authority, the Society of Apothecaries. Elizabeth Garrett Anderson had obtained the Licentiate of the Society in 1865, becoming the first woman to do so. Persisting with their intolerant and sexist attitude to women, the Society had sought legal advice to stop Anderson taking the exam but were compelled to allow her. Shortly after Hoggan passed the preliminary exams herself in 1867, the Society's lawyers closed the loophole and they too formally excluded women from its professional examinations.

Not to be deterred in any way, Frances Hoggan's progress was swift and sure: at the University of Zurich, Morgan completed the medical course in three years rather than the usual five, and in March 1870, became only the second woman to gain an MD, with a thesis on progressive muscular atrophy at Zurich University. She was now the first British woman to obtain a European MD degree. Afterwards, at a Vienna clinic she completed a course on operative midwifery as a pupil of surgeon Gustav Braun.

Like others, she spent a number of years as a medical practitioner working with Elizabeth Garrett Anderson at the New Hospital for Women; she also helped to found the National Health Society with Elizabeth Blackwell in 1871.

In 1874, she married Dr George Hoggan and obtained her licence to practice in the UK from the King's and Queen's College of Physicians of Ireland in February 1877.

In 1882, she called for a publicly funded women's medical service for female patients in India which led the way for the Dufferin Fund.[1] She also became medical superintendent at the North London Collegiate School, one of the country's first rigorously academic secondary schools for girls.

Her 1884 paper, 'The Position of the Mother of the Family', made use of the latest understanding about conception and reproduction to argue that mothers should have more rights over their children.[2]

Frances and her husband George wrote a number of papers together and were anti-vivisectionists.[3] In an article for the *Vaccination Inquirer* in September 1883, they both argued against compulsory vaccination.

Hoggan became a campaigner and social reformer, and toured the United States lecturing. She had a particular interest in racial issues, and was a speaker at the Universal Race Congress in London in 1911.

Notes

1. Frances Hoggan, *Medical Women for India*, Bristol, 1882.
2. Susan G. Bell and Karen M. Offen, *Women, the Family, and Freedom: 1880–1950*, Stanford University Press, 1983.
3. For example, George Hoggan and Frances Elizabeth Hoggan, 'On the Comparative Anatomy of the Lymphatics of the Uterus', *Journal of Anatomy and Physiology*, 16 (Pt 1), 50–89, 1881.

Edith Pechey (1845–1908)

Mary Edith Pechey spent time in India; she was a keen campaigner for women's rights and was active in a wide range of social issues. After being educated at home by her parents, she took work as a governess and teacher until 1869. Lutzker notes that 'Her mother also was competent in Ancient Greek and other studies and both parents possessed – along with their questing nonconformist minds – a deep and serious love of learning.'[1]

Responding to Sophia Jex-Blake's famous advertisement in *The Scotsman*, Pechey became one of the Edinburgh Seven, proving her ability by winning the top grade in the chemistry exam in her first year, making her eligible for a Hope Scholarship.[2]

However, male insecurity and spineless self-interest and sexism reared their ugly heads again when Crum Brown, the then Professor of Chemistry was, according to *The Spectator*:

> concerned that awarding the scholarship to a woman would provoke a backlash from the male students, who had grown increasingly hostile when they saw that women were capable of outstripping them in competitive examinations. He had also noticed that, as the term had progressed, many of his respected colleagues in the medical faculty began to express resentment of the presence of women in the University.

He therefore decided to award the scholarships to male students who achieved lower grades than Pechey. The cited reason was 'that women are not part of the University class, because they are separately taught'.[3]

Things just kept getting worse while the officials fumbled ineptly around. According to Shirley Roberts:

> Having used the matter of separate classes as a reason not to award the scholarship to Pechey, Crum Brown felt unable to issue the women with the usual certificates of attendance at his chemistry classes. Instead, he gave them credit for attending a 'ladies' class in the University. Only the standard certificates met the Faculty's requirements for the medical degree; the Professor's 'Strawberry Jam Labels', as Sophia Jex-Blake referred to them were useless.[4]

Jex-Blake records that the women appealed to the Senatus Academicus. Edith Pechey stated her claim to a scholarship and the other women asked that they be granted the standard certificates for their chemistry classes. The Senatus met on 9 April 1870 and, after some debate, ruled in favour of the women on the certificates but against them on the Hope Scholarship.[5]

The press generally had a field day at the expense of the older academics at Edinburgh with their ancient values and insecurities:

The Times:

> [Miss Pechey] has done her sex a service, not only by vindicating their intellectual ability in an open competition with men, but still more by the temper and courtesy with which she meets her disappointments.

The Spectator was waspish:

> To make women attend a separate class, for which they have to pay, we believe, much higher fees than usual, and then argue that they are out of the pale of

> competition because they do so, is, indeed, too like the captious schoolmaster who first sent a boy into the corner and then whipped him for not being in his seat.

This is what a frustrated Pechey did next. She contacted the College of Physicians in Ireland to let her take exams, leading to a licence in midwifery. The College had begun licensing women doctors; Pechey passed its exams in Dublin in May 1877. She worked for a time at the Birmingham and Midland Hospital for Women, despite the lack of an official qualification, taking over from Louisa Atkins.

Next she went to the University of Bern, passed her medical exams in German at the end of January 1877 and was awarded an MD with a thesis 'Upon the constitutional causes of uterine catarrh'. For the next six years, Pechey practised in Leeds, covering women's health education and lecturing in a number of medical areas including nursing.

Partly as a reaction to the exclusion of women by the International Medical Congress, she established the Medical Women's Federation of England and in 1882 was elected president. The Medical Women's Federation (or Foundation) built upon the Association of Registered Medical Women, which had been founded in London in 1879 with the intention that it would 'speak on behalf of all medical women and represent their interests'.[7]

Notes

1. Edythe Lutzker (1967), 'Edith Pechey-Phipson, M.D.: untold story', *Medical History*, 11(1), 41–45.
2. Shirley Roberts, *Sophia Jex-Blake: A Woman Pioneer in Nineteenth-Century Medical Reform* (2015 ed.), New York, 1993, pp. 86–87.
3. *The Spectator*, 9 April 1870.
4. Shirley Roberts, *Sophia Jex-Blake*, p. 92.
5. Sophia Jex-Blake, *Medical Women: A Thesis and History*, Edinburgh: Oliphant, Anderson & Ferrier, pp. 78–83, 1886.
6. The International Medical Congress was a series of international scientific conferences on medicine that took place, periodically, from 1867 until 1913. The Medical Women's Federation is today the largest UK body of women doctors. The organisation is dedicated to the advancement of the personal and professional development of women in medicine and to improving the health of women and their families in society.
7. Matt Elsom, 'Breaking the Mould: The Entry of Women into Medicine in the UK', 2011; Helen Jones, *Women in British Public Life, 1914–50: Gender, Power and Social Policy*, London, 2014.

Dame Mary Scharlieb (1845–1930)

As an Indian woman, Mary Scharlieb overcame what would seem to be insurmountable obstacles to achieve what she did. The Wellcome Collection tells us how Mary Ann Dacomb Scharlieb (née Bird) was one of the first generation of British women doctors, from the inaugural class of students at the London School of Medicine for Women. By then she was already married and a mother.

Scharlieb's objective in acquiring a medical education (she had already obtained a medical qualification at the Madras Medical School) was primarily to alleviate the unnecessary sufferings of purdah women in India who could not be treated by male doctors, by bringing to them the benefits of modern Western medicine in place of native midwives. Returning to London in 1887, she embarked on a long and distinguished career, was recognised as an accomplished gynaecological surgeon, and played a busy part in public life: she sat on the Royal Commission on Venereal Diseases 1913–1916 and was later a leading figure of the Social Purity movement.[1]

Here is a small selection of her many achievements:

- After some predictable opposition from the authorities, she was allowed to attend the Lying-in Hospital daily; she wished then and there to join the pioneer women in England in their struggle for London degrees.
- She became one of the first four women students at the Madras Medical School and in three years she gained her Licentiate in Medicine. Burdened with chronic ill health, she soldiered on and with her children in 1878 she returned to England where she was reluctantly taken on by Mrs Garrett Anderson, who had recently started the London School of Medicine for Women. In November 1882, at the age of 37, she qualified in medicine with honours in all subjects, the Gold Medal and the Scholarship in Obstetrics; shortly after this she gained second-class honours in Surgery followed by an elective in Vienna.
- With the help of Sir Henry Acland, she was commanded to Windsor, 'where Queen Victoria insisted on hearing fully from her how the women of India, forbidden by their religion to obtain help from male doctors, were abandoned to the ignorance and superstition of the native women who attended them in childbirth. The Queen (Victoria), always mindful of her Indian subjects, expressed her warm approval and interest to Mrs. Scharlieb,

and waxed indignant that others had tried to persuade her that there was no need for medical women in India.'[2]
- Obtained her MD from the University of London in 1888. From 1887 to 1902 she was surgeon at the New Hospital for Women (later the Elizabeth Garrett Anderson Hospital), initially assisting Dr Elizabeth Garrett Anderson, and being senior surgeon from 1889.
- In 1887, she was appointed lecturer on forensic medicine to the Royal Free Hospital, in 1889 lecturer on midwifery (until 1913), and in 1902 chief gynaecologist. In 1897, she had obtained the Master of Surgery degree, making her the first woman in Britain to secure the MD and the MS degrees.[3]
- She was a member of the Council of Obstetrics and Gynaecological Section of the Royal Society of Medicine, but as a woman was barred from participating in the running of the society or the election of officers.

Mary Scharlieb continued her pioneering and distinguished career well into the twentieth century with forty or so years' practice in Harley Street. An important theme running through Scharlieb's writings was the importance of including a female point of view, in both medical and legislative arenas. This principle was also exemplified in her commitment to providing medical care for women in India.[4]

Notes

1. https://wellcomecollection.org/works/aw5shdew
2. https://anglicanhistory.org/bios/mscharlieb.html; London: The Catholic Literature Association, 1933.
3. Michele Stokes (2004), *A measure of the elite: a history of medical practitioners in Harley Street, 1845–1914* (PhD). University of London.
4. Greta Jones, 'Women and eugenics in Britain: The case of Mary Scharlieb, Elizabeth Sloan Chesser, and Stella Browne', *Annals of* Science, 52(5), 481–502, 1995.

Matilda Chaplin Ayrton (1846–1883)

Matilda Ayrton's journey through the quagmire that came to characterise qualification in and recognition for a medical degree permitting her to practise as a doctor is of course very similar to the tortuous circuitous journeys of those who went before her.

She was born into a middle-class professional family in which the education of the children was paramount. Matilda began her study of medicine in

1867, spending two years at the Ladies' Medical College, and passing the preliminary examination at Apothecaries' Hall in 1869; she was refused admission for the higher exam – because she was a woman. She was one of the defiant 'Seven against Edinburgh' women and although she eventually enrolled at the University of Edinburgh, she was barred from lectures in medicine. However, she was able, by resorting to law, to gain honours in anatomy and surgery at the extramural examinations held in 1870 and 1871 at Surgeons' Hall, Edinburgh, before an 1872 judgement finally prohibited women students.[1]

In 1871, with the doors of medical classes in England and Scotland firmly closed to her, she did what others had done before her – she completed her education at Sorbonne University Paris, where every facility was afforded her. The University recognised her abilities by bestowing upon her the degrees of Bachelier ès-Sciences and Bachelier ès-Lettres.

Early the next year, she obtained a certificate in midwifery from the London Obstetric Society, the one and only medical qualification then obtainable by women in England.

The Obstetrical Society of London

The Obstetrical Society of London was formed in 1858 and merged in 1907 with the Royal Medical and Chirurgical Society of London to form the Royal Society of Medicine (RSM).

The Society was set up in 1858, the successor to an Obstetric Society dating from 1825, and in the aftermath of the Medical Act 1858.[2] Over its first fifteen years, the membership of the Society rose to about 600.

The Act's proposals included regulation of medical practitioners, taken at the time to include midwifery; and the Society turned in time to certifying midwives. The diploma introduced in 1872 recognised the role of the midwife, in supervising 'normal labour' and commonly appeared in advertisements or curriculum vitae as *Cert., L.O.S.* Training was provided at London hospitals such as the City of London Lying-in Hospital.

While in Japan with her husband William Edward Ayrton, whom she married in 1872, Matilda pursued her interest in anthropology research and opened a school for Japanese midwives, in which she lectured with the aid of an interpreter.

In 1877, she developed tuberculosis and returned to Europe where two years later she took her MD at Paris, and presented as her thesis the results of her Japanese studies entitled *Recherches sur les dimensions générales et sur le développement du corps chez les Japonais* (Paris, 1879). In 1880, she became a licentiate of the King and Queen's College of Physicians in Ireland, where she was the only female candidate at the examinations heading the pass lists in medicine and midwifery. In 1880, she lived in London, her time largely taken up with studying diseases of the eye at the Royal Free Hospital.

The Ladies' Medical College and the Female Medical Society

The Ladies' Medical College (renamed the Obstetrical College for Women) was a short-lived English medical college for women, established in 1864 in Fitzroy Square by the Female Medical Society. It offered courses in midwifery and gynaecologic and paediatric diseases but, when it failed to extend its curriculum to the full range of medical training, it closed in 1873.

The Medical Act 1858 created the General Medical Council to regulate doctors in the UK. The Act notes that 'it is expedient that Persons requiring Medical Aid should be enabled to distinguish qualified from unqualified Practitioners'. The Act effectively barred women from becoming doctors in the UK by requiring physicians to pass examinations offered by any of nineteen examination boards, none of which admitted women. Midwifery, however, was unregulated, and was undertaken either by self-trained working-class women, or by male physicians.

In an attempt to swerve round this, the Female Medical Society was established in 1862, an aim of which was to enable women to qualify as doctors, but with a focus on medicine applicable to women and children. The society also aimed to elevate the status of midwifery to a profession for educated women, not least to cater for the feelings of female patients who preferred to be seen by female practitioners; and to save lives and reduce perinatal deaths, both through better training and practice, but also by reducing infections introduced by male physicians through their routine contact with disease, surgery and post-mortems.

From 1867, students gained their clinical experience at the British Lying-In Hospital (est. 1749) in Long Acre where there were eighty-four students by 1870. In 1849 it relocated to Endell Street in a forty-bed hospital. Notable students of the Ladies Medical College include Isabel Thorne (who described its teaching as inadequate) and Matilda Chaplin, both of whom went on to be members of the Edinburgh Seven; Florence Fenwick Miller and Alice Vickery, the first British woman to qualify as a chemist and pharmacist.

Notes

1. Joan Mason, 'Hertha Ayrton (1854–1923) and the Admission of Women to the Royal Society of London', *Notes and Records of the Royal Society of London*, 45(2), 1991, 201–220.
2. Philip K. Wilson, *Childbirth: The Medicalization of Obstetrics*, Taylor & Francis, 1996, pp. 38–39.

Mary Murdoch (1864–1916)

Mary Murdoch was a Scottish-born physician and suffragist who had a lifelong association with Hull where she was the first woman doctor, a suffragist, and flamboyant fast car owner which she drove 'at speed'.

She lived in Park Street, close to the grand Grade II-listed former Victoria Hospital for Sick Children. Opened in 1890, it was where Hull's first female GP (from 1910) Dr Murdoch worked as a house surgeon in her first medical appointment in 1893 after graduating from the London School of Medicine for Women. In 1900, Mary appointed Dr Louisa Martindale as her assistant, and they lived together at Grosvenor House on Beverley Road.

Murdoch was a passionate believer that women should be able to be seen by a woman doctor if that's what they wanted. In an address she gave to students at the London School of Medicine for Women in 1904, she said that she dreamt of a future in which it would be seen as 'One of the barbarisms of a past age that a medical man should ever have attended a woman'.

She was way ahead of her time – she founded the first crèche in Hull as well as a school for mothers.

The former hospital (now Victoria House, accommodating 128 flats) was replaced by Hull Royal Infirmary.

Elsie Inglis (1864–1917)

Eliza Maud 'Elsie' Inglis was a Scottish medical doctor, surgeon, teacher, suffragist, and founder of the Scottish Women's Hospitals. She was the first woman to hold the Serbian Order of the White Eagle.

Inglis was born in the hill station town of Naini Tal, British India, one of eight; unusually, her parents considered the education of a daughter as important as that of a son,[1] and had them schooled in India. Elsie and her sister Eva owned forty dolls which she used to treat for the 'spots' (measles) she had painted on.[2]

In 1887, the Edinburgh School of Medicine for Women was opened by Dr Sophia Jex-Blake; Inglis started there. As a reaction to Jex-Blake's methods, and after two fellow students Grace and Georgina Cadell were expelled, Inglis and her father founded the Edinburgh College of Medicine for Women, under the auspices of the Scottish Association for the Medical Education of Women, whose sponsors included Sir William Muir, Principal of the University of Edinburgh. Inglis's sponsors also arranged clinical training for female students under Sir William MacEwen at the Glasgow Royal Infirmary.

In 1892, she obtained the Triple Qualification, becoming a Licentiate of the Royal College of Physicians of Edinburgh, the Royal College of Surgeons of Edinburgh and the Faculty of Physicians and Surgeons of Glasgow. She was concerned at the low standard of care and lack of specialisation in the needs of female patients, and so obtained a post at Elizabeth Garrett Anderson's pioneering New Hospital for Women in London, and then at the Rotunda in Dublin, a leading maternity hospital. Inglis gained her MBChM qualification in 1899, from the University of Edinburgh, after it opened its medical courses to women.

Inglis returned to Edinburgh in 1894, completed her medical degree and became a lecturer in gynaecology at the Medical College for Women and then set up a medical practice. Given that women and children's medicine was under resourced, she opened a maternity hospital, named The Hospice, for poor women alongside a midwifery resource and training centre, initially in George Square. The Hospice gained an accident and general service as well as maternity,[3] with an operating theatre and eight beds[4] on the Royal Mile,[5] and was the forerunner of the Elsie Inglis Memorial Maternity Hospital. In 1913, Inglis travelled to Michigan to visit and learn from a new type of maternity hospital.

Inglis often waived the fees owed to her and would pay for her patients to recuperate by the seaside, with polio being one of the childhood illnesses she was particularly concerned with. She was a consultant at Bruntsfield Hospital, a nearby hospital for women and children.

Inglis was in a relationship for some time with Flora Murray, a fellow doctor and suffragette.

Suffrage

She was the secretary of the Edinburgh National Society for Women's Suffrage in the 1890s while she was studying her medical degree.

Inglis was honorary secretary from the early years of the Scottish Federation of Women's Suffrage Societies from 1906 until 1914. She spoke in 1907 on a platform with Jessie Chrystal Macmillan and Alice Low as fellow speakers.[7]

Notes

1. Hamish MacPherson (5 May 2020), 'Greatest Scot? The many talents of Dr Elsie Inglis', *The National*, p. 20: 'when she saw suffering or injustice she wanted to make a difference ... and what she did paved the way for other women to come after her'. Alan Cumming in an interview with Nan Spowart on 11 November 2017, *The National.*
2. William Knox, *Lives of Scottish Women: Women and Scottish Society, 1800–1980*, Edinburgh, 2006, p. 141.
3. Lucy Inglis (8 November 2014), 'Elsie Inglis, the suffragette physician', *The Lancet*, 384 (9955), 1664–1665.
4. Frances Balfour, *Dr. Elsie Inglis*, New York, p. 182, 1919.
5. 'The legacy of Elsie Inglis – Edinburgh's shame', The History Company, 6 February 2014.
6. Margot Lawrence, *Shadow of Swords: A Biography of Elsie Inglis*, Northumberland Press Limited, p. 81, 1971.
7. The Chrystal Macmillan Prize is a £100 award given 'at the discretion of the Scholarships and Prizes Committee' of the Honourable Society of the Middle Temple in London, a professional group of attorneys. The prize was founded as an annual grant to benefit female law students scoring the highest in the bar's final examination, and to support societies with which Chrystal Macmillan was associated.

Mona Chalmers Watson (1872–1936)

At this time, women studying medicine risked social isolation, 'contemptuous glances' and 'sneering remarks'.[1]

Mona Chalmers Watson was head of the Women's Army Auxiliary Corps. The first woman to receive an MD from the University of Edinburgh in 1896, she helped found the Elsie Inglis Hospital for Women, was the first president of the Edinburgh Women's Citizen Association and senior physician at the

Edinburgh Hospital and Dispensary for Women and Children. Through her mother's brother, Mona was related to Elizabeth Garrett Anderson.

Chalmers Watson then spent a year in London working as a physician at the Maternity District Association at Plaistow followed by six months working at Dr Barnardo's Homes in Kent.

Her experience in Plaistow, where her record consisted of more than 1,000 confinements with a mortality rate of just over 1 per 1,000,not surprisingly informed her MD thesis when she returned to Edinburgh the following year.[2]

Notes

1. 'Lives and Times – Number 109 Alexandra (Mona) Chalmers (1872–1936)', *The Scotsman*, 8 September 2005, pp. S2 37.
2. Geddes Campbell (1898), Alexandra Mary, 'A year's obstetric work at Plaistow Maternity Institute, London: statistics and notes on fifteen hundred and fifty cases, with observations on cases of chorea gravidarum, extra uterine gestation, rupture of the uterus, and inversion of the uterus.'

Elizabeth Garrett Anderson (1873–1943)

> '[A] doctor leads two lives, the professional and the private, and the boundaries between the two are never traversed.'
>
> —Elizabeth Garret Anderson

We have touched a number of times already on Elizabeth Garrett Anderson's defiant and pioneering lifework in getting women accepted as medical doctors in the face of male sexism and professional insecurity. Having been rejected by the men-only medical schools, she turned her fire on the Worshipful Society of Apothecaries, seeking permission to sit their Licentiate examination, which would enable her to register as a medical practitioner.

The Society informs us that a letter to the Royal College of Physicians in 1864 shows that she viewed it as 'a comparatively easy mode of entrance' and that she believed instead that 'in the interest of the public and for the honour of the profession women should be subject to the severest tests open to men'. However, it was her only available option and she pieced together private unofficial study, lecture attendance at a variety of institutions, and clinical practice at the Middlesex Hospital in order to fulfil the requirements for the exam.[1]

To the Apothecaries, this was a preposterous situation which they were not going to take lying down. They took recourse to law in an attempt to exclude Garret Anderson from their examination but were unable to refuse her under the terms of their own Charter.

She passed the Licentiate of the Society of Apothecaries (LSA) exam on 28 September 1865 and was therefore eligible and free to practise as a doctor. Despite the LSA, as a woman she still could not hold a medical post in any British hospital. So in late 1865, Garrett opened her own practice at 20 Upper Berkeley Street. During the 1865 cholera epidemic, Garrett set up St Mary's Dispensary for Women and Children at 69 Seymour Place. In the first year, she tended to 3,000 new patients, who made 9,300 outpatient visits to the dispensary. In 1872, the dispensary became the New Hospital for Women and Children, treating women from all over London for gynaecological conditions; the hospital moved to Marylebone Street in 1874.

Garrett Anderson was the first woman qualified in Britain to do so openly; before that there was Dr James Barry who was born and raised female but presented as male from the age of 20.[2] The Society of Apothecaries immediately amended its regulations to prevent other women obtaining a licence, meaning that Jex-Blake, for example, could not follow this same path; the new rule prohibited privately educated women from being eligible for examination.

In 1870, Garrett Anderson became the first woman to graduate with a medical degree from Sorbonne University. She had learnt French to facilitate this when she learnt the dean of the medical faculty was open to admitting women students. In addition, she was, of course, a co-founder and dean of the London School of Medicine for Women, which was the first medical school in Britain to train women as doctors.

She was made a visiting physician of the East London Hospital for Children (later the Queen Elizabeth Hospital for Children), becoming the first woman in Britain to be appointed to a medical post. As such, she was thus the first female dean of a British medical school. She was the first woman in Britain to be elected to a school board (The London School Board) and, as mayor of Aldeburgh, the first female mayor in Britain.

Before all that, she had continued her struggle to qualify by studying privately with various professors, including some at St Andrews, the

Edinburgh Royal Maternity and the London Hospital Medical School. In 1860, after snubs from Harley Street doctors, Garrett spent six successful months as a surgery nurse at Middlesex Hospital. She was permitted to attend an outpatients' clinic, and then perform her first operation. She unsuccessfully attempted to enrol in the hospital's medical school but was allowed to attend private tuition in Latin, Greek and pharmacology with the hospital's apothecary, while continuing her work as a nurse. She took on a tutor in anatomy and physiology three evenings a week. Eventually she was allowed into the dissecting room and the chemistry lectures.

All the while, Garrett was *persona non grata* among the chauvinist male students, who in 1861 presented a memorial to the school against her admittance as a fellow student, despite the support she enjoyed from the administration. She was forced to leave the Middlesex Hospital but not without first obtaining an honours certificate in chemistry and materia medica.

Garrett then applied to several medical schools, including Oxford, Cambridge, Glasgow, Edinburgh, St Andrews and the Royal College of Surgeons, all of which refused her admittance.

Underpinning it all was the vexatious discourse regarding male views on women in medicine. In 1874, Henry Maudsley's article on *Sex and Mind in Education* was published, arguing that education for women caused them over-exertion and thus reduced their reproductive capacity, sometimes causing 'nervous and even mental disorders'. Garrett counter-argued that the real peril for women was not education but boredom and that fresh air and exercise were better for them than sitting by the fire with a novel.[3]

The bastion that was the BMA

In 1873, Garrett attained membership of the British Medical Association (BMA). Five years later, in 1878 (male bastions move slowly), a motion was proposed to exclude women following the election of Garrett Anderson and Frances Hoggan. The motion was opposed by Dr Norman Kerr, who maintained the equal rights of members.[4]

M. A. Elston points out that this was 'one of several instances where Garrett, uniquely, was able to enter a hitherto all male medical institution which subsequently moved formally to exclude any women who might seek to follow her'.[5] In 1892, women were again admitted to the British Medical

Association following a lengthy campaign by Anderson and others. In 1897, Garrett Anderson was elected president of the East Anglian branch of the BMA.

The New Hospital for Women was a fine building in the Euston Road; for many years, the hospital was staffed exclusively by medical women. The schools (in Hunter Street, WC1) had over 200 students, most of them preparing for the medical degree of London University, which was opened to women in 1877.

Suffrage

Garrett Anderson also found time to actively support the suffrage movement. In 1866, Garrett Anderson joined the first British Women's Suffrage Committee, although her sister, Millicent Garrett Fawcett, eclipsed her in terms of action. Garrett Anderson became a member of the Central Committee of the National Society for Women's Suffrage in 1889. Her daughter Louisa, also a doctor, was more active and more militant, spending time in prison in 1912 for her suffrage activities.

Notes

1. https://historycollections.blogs.sas.ac.uk/2018/07/09/womens-history-at-the-worshipful-society-of-apothecaries
2. James Barry, a woman who lived as a man, qualified before her; see above. Elizabeth Blackwell was the first woman registered by the General Medical Council in the UK, having qualified in the United States. Frances Hoggan preceded Anderson as the first British woman to qualify in Europe, having qualified in Switzerland.
3. He of Maudsley Hospital fame, and the annual Maudsley Lecture of the Royal College of Psychiatrists.
4. *The Times*, 9 August 1878, p. 7.
5. M. A. Elston, 'Anderson, Elizabeth Garrett (1836–1917), physician', *Oxford Dictionary of National Biography* (online ed.), Oxford University Press, 2005.

Nursing

'...it never entered my head that a nurse would be capable of taking and recording temperatures.'

—Dr Francis de Haviland Hall, Physician to Westminster Hospital, 1894–1912, writing in 1899

Mary Seacole (1805–1881)

The inane quotation above by Dr Hall says a lot more about the arrogance and sexism exhibited by some of the male medical establishment than it does about the accomplishments of the nursing profession at the time. Mary Jane Seacole is an example of the professionalism and competency of our nurses. She is remembered as an outstanding British nurse and businesswoman, famous for her nursing work during the Crimean War (1853–1856) and for publishing the first autobiography written by a black woman in Britain. She was the daughter of James Grant, a Scottish Lieutenant in the British Army.

Her mother, Mrs Grant, nicknamed 'The Doctress', was a healer who dabbled in traditional Caribbean and African herbal medicines. Mrs Grant was also landlady at Blundell Hall, a highly respectable boarding house at 7 East Street, Kingston.

In the eighteenth century, Jamaican doctresses, proficient as they were in folk medicine, included in their armamentaria the routine use of hygiene and herbs. They had an impressive knowledge of tropical and infectious diseases, and could rival any general practitioner in the treatment of ailments and injuries; this was all acquired from their time spent tending fellow slaves on sugar plantations.[1] At Blundell Hall, Seacole honed her nursing skills. Blundell Hall doubled as a convalescent home for military and naval personnel recuperating from illnesses such as cholera and yellow fever.

Seacole's autobiography engagingly relates how she began experimenting in medicine, based on what she learned from her mother, by practising on a doll and then progressing to pets before assisting her mother in treating human patients. Because of her family's close association with the military, she was able to pick up some military medicine, combining this with the West African remedies she got from her mother. In the late eighteenth and early nineteenth century, Jamaica neonatal deaths exceeded 25 per cent of all births. However, Seacole, using traditional West African herbal remedies, claimed that she never lost a mother or her child.

She treated and nursed patients through the cholera epidemic of 1850, which carried away some 32,000 Jamaicans. In 1853, Mary was in Kingston, caring for victims of a yellow fever epidemic.[2] She was invited by the medical authorities to supervise nursing services at the British Army's headquarters, and she repurposed New Blundell Hall, her mother's former lodging house rebuilt after a fire, to operate as a purpose-built hospital.

The Crimean War was raging: Seacole spotted a business opportunity here and so she travelled to England; the British War Office declined her offer to be sent as an army nurse to the Crimea, even though medical facilities for wounded soldiers were notoriously appalling.

Mary was not to be denied: she funded her own trip to the Crimea with plans to open 'The British Hotel' in Balaclava, more a restaurant than a hotel which catered for up to fifty diners and offered the services of a barber. She had arrived at Balaklava on 19 March 1855 on board *Albatross*, a cattle transport, which, according to Margrave, she boarded in Constantinople after her voyage from England on *Hollander*.

The ship put in at Malta, where Seacole encountered a doctor who had recently left Scutari. He wrote for her a letter of introduction to Florence Nightingale. Seacole called on Nightingale at the Barrack Hospital in Scutari, where she requested a bed for the night: she wrote of a rather cagey Selina Bracebridge, an assistant to Nightingale:

> Mrs. B. questions me very kindly, but with the same look of curiosity and surprise. What object has Mrs. Seacole in coming out? This is the purport of her questions. And I say, frankly, to be of use somewhere; for other considerations I had not, until necessity forced them upon me. Willingly, had they accepted me, I would have worked for the wounded, in return for bread and water. I fancy Mrs. B— thought that I sought for employment at Scutari, for she said, very kindly – 'Miss Nightingale has the entire management of our hospital staff...'

Seacole informed Bracebridge that she intended to travel to Balaclava the next day to join her business partner. She reports that her meeting with Nightingale was friendly, with Nightingale asking, 'What do you want, Mrs Seacole? Anything we can do for you? If it lies in my power, I shall be very happy.'

A bed was then found for her and breakfast served in the morning, with a 'kind message' from Bracebridge. A footnote in the memoir states that Seacole subsequently 'saw much of Miss Nightingale at Balaclava', but history records no further meetings.

Seacole embarked on the four-day voyage to the British bridgehead at Balaclava. In the absence of viable building materials, Seacole gathered abandoned detritus, largely metal, glass doors and window-frames and driftwood, with which to build her hotel on the main British supply road from Balaclava to the British camp near Sevastopol, and within a mile of the British headquarters.

She opened in March 1855. An early and helpful visitor was Alexis Soyer, the famous French chef who had come to Crimea to help improve the dreadful diet of British soldiers. He records meeting Seacole in his 1857 work *A Culinary Campaign* and advised her to focus on a food and beverage service, and not to bother with beds for visitors because they either slept on ships in the harbour or in tents in the camp.[3]

The building was provisioned from London and Constantinople, and local purchases from the British camp near Kadikoi and the French camp at nearby Kamiesch. Seacole records how she sold anything – 'from a needle to an anchor' – to army officers and visiting sightseers. Meals were of course served at the hotel, cooked by two black cooks, the kitchen also offered outside catering. Despite persistent thefts of livestock, Seacole's establishment prospered.

Seacole was grateful for the kind hospitality Nightingale had given her; Seacole has pointed out that when she tried to join Nightingale's group, one of Nightingale's colleagues rebuffed her – suggesting that racism was behind that rebuttal. On the face of it, Nightingale was gracious and complimentary, acknowledging that 'she has done a deal of good for the poor soldiers', but private correspondence to her brother-in-law reveals her to have been somewhat catty and deceitful:

> I had the greatest difficulty in repelling Mrs Seacole's advances, and in preventing association between her and my nurses (absolutely out of the question!) ... Anyone who employs Mrs Seacole will introduce much kindness – also much drunkenness and improper conduct.

Seacole also provided catering for spectators at the battles, and spent time on Cathcart's Hill, some 3.5 miles north of the British Hotel. In his dispatch of 14 September 1855, William Howard Russell, special correspondent of *The Times*, reported that she was a 'warm and successful physician, who doctors and cures all manner of men with extraordinary success. She is always in attendance near the battlefield to aid the wounded and has earned many a poor fellow's blessing.' Russell also told readers that she 'redeemed the name of sutler'. Seacole habitually wore vivid and flamboyant clothing.

Her peers, though initially cautious, soon began to realise what an asset Mary was to the nursing assistance afforded to the wounded and to morale in general. One British medical officer described Seacole in his memoir as

> The acquaintance of a celebrated person, Mrs. Seacole, a coloured woman who out of the goodness of her heart and at her own expense, supplied hot tea to the poor sufferers [wounded men being transported from the peninsula to the hospital at Scutari] while they are waiting to be lifted into the boats ... She did not spare herself if she could do any good to the suffering soldiers. In rain and snow, in storm and tempest, day after day she was at her self-chosen post with her stove and kettle, in any shelter she could find, brewing tea for all who wanted it, and they were many. Sometimes more than 200 sick would be embarked in one day, but Mrs. Seacole was always equal to the occasion.

But Seacole went further, often carrying bags of lint, bandages, needles and thread for the soldiers' wounds.

In late August, Seacole won a bet that she would be the first British woman to enter Sevastopol after it fell. Pass in hand, she toured the shattered town, visiting the crowded hospital by the docks, containing many dead and dying Russians.

The Treaty of Paris was signed on 30 March 1856, after which the soldiers left Crimea. Financially Seacole was in a precarious position, overstocked as she was with unsaleable provisions, and more goods arriving daily, with creditors demanding payment. She had no choice but to auction as much as possible at below-cost prices before the soldiers left for their homes. Seacole was one of the last to leave the Crimea, returning to England 'poorer than [she] left it'. Though she left poorer, there was some small consolation in that her impact on the soldiers' morale was invaluable to the soldiers she treated, as described in *The Illustrated London News*:

> Perhaps at first the authorities looked askant at the woman-volunteer; but they soon found her worth and utility; and from that time until the British army left the Crimea, Mother Seacole was a household word in the camp … In her store on Spring Hill she attended many patients, cared for many sick, and earned the good will and gratitude of hundreds.[4]

In short, Mary Seacole, although never the 'black British nurse' she is often claimed to have been, was a successful, trailblazing mixed-race immigrant to Britain. She led a full, adventurous life.

Seacole was bankrupt when she returned to London, destitute and in failing health. Queen Victoria's nephew Count Gleichen had become a friend of Seacole's in Crimea. He supported fundraising efforts on her behalf.

Mrs Seacole came late to the war because she had flown in from Panama and was busy in London organising her gold investments – and where she provided services for prospectors going overland to and from the California Gold Rush.

She was largely forgotten for almost a century after her death. Her autobiography, *Wonderful Adventures of Mrs. Seacole in Many Lands* (1857), is the first autobiography written by a black woman in Britain. In 2017, Robert McCrum chose it as one of the 100 best nonfiction books, calling it 'gloriously entertaining'.[5]

Notes

1. Moira Ferguson, *Nine Black Women*, London, 1998, p. 68; Jane Robinson, *Mary Seacole: The Charismatic Black Nurse who Became a Heroine of the Crimea*, Constable, 2004.
2. Paul Chrystal, *A History of the World in 100 Pandemics, Plagues and Epidemics*, Pen & Sword, Barnsley, 2023.
3. In 1855, reports of the conditions, described as 'appalling privations', endured by British soldiers in the Crimean War were causing outrage in the press. Having read in *The Times* of what Cowen calls 'the privations being suffered by the troops', who were 'racked by

disease and near-starvation', Alexis Soyer offered the government his services, free of charge. The offer was accepted and he went out to Scutari and Constantinople, transforming the catering in the hospitals there. He then accompanied Florence Nightingale to Balaklava and Sevastopol and upgraded the provisioning of the field hospitals, in addition to taking on the cooking for the Fourth Division of the army.

4. Jane, Robinson, *Mary Seacole: The Charismatic Black Nurse Who Became a Heroine of the Crimea*, Constable, 2004, p. 153, quoting *The Illustrated London News*, 30 August 1856.
5. Robert McCrum (10 April 2017), 'The 100 best nonfiction books: No. 62 – *The Wonderful Adventures of Mrs Seacole in Many Lands* (1857)', *The Guardian*.

Florence Nightingale (1820–1910)

> ... in that house of misery
> A lady with a lamp I see
> Pass through the glimmering gloom,
> And flit from room to room.
>
> —Henry Wadsworth Longfellow,
> *Santa Filomena*, 1857

Nightingale's supreme achievement in lifting nursing out of the doldrums which characterised the job as a form of labour into a much more pro-active, positive profession are well documented. Suffice for us to summarise the major changes she made, starting from her experiences in Crimea which brought nursing much closer to and much more recognisable to the profession we know today.

Hygiene and infection control were central, always, in every aspect and scenario of nursing care (starting with simple but super-effective handwashing and Nightingale Wards).[1] An holistic approach to nursing care was essential. She recognised that hospital and ward design were influential in assisting and expediting recovery and well-being, as well as the importance of a healthy diet.

She was invited by the then Minister of War, Sidney Herbert, to recruit women nurses into military hospitals in the Crimea to counter the shocking conditions endured by injured British and other troops. She and the thirty-eight nurses who went with her plus fifteen Catholic nuns (the Sisters of Mercy) improved supplies of beds, blankets and food in the hospitals, as well as the general conditions and cleanliness at Scutari. Morale must have improved immeasurably as well, aiding recovery and a return to the front line wherever possible.

In the middle of it all, Nightingale sent a plea to *The Times* for government intervention over the dire facilities; the government commissioned Brunel to design a prefabricated hospital that could be built in England and shipped out to the Dardanelles. This became Renkioi Hospital, a civilian facility that had a mortality rate less than one tenth of that of Scutari.

Her social reforms included improving healthcare and sanitation for all sections of British society but particularly in working-class housing, advocating better hunger relief in India, helping to abolish prostitution laws that were injurious to women, and expanding female participation in the workforce.

Nightingale was a pioneer in the use of data and statistics in nursing as well as in presentation, deploying as she did skills and technique in a uniquely visual way to make things easy to understand (infographics and statistical graphics).

> Her obsession for codification and collection and classification of data and information began during her travels abroad and she was the first woman to be elected to the Royal Statistical Society. She was constantly innovative in the way she presented her arguments visually and is credited with developing a form of pie-chart now known as the polar area diagram.

Edith Cavell (1865–1915)

> Standing as I do in view of God and Eternity, I realise that patriotism is not enough, I must have no hatred or bitterness towards anyone.
>
> —Edith Cavell, speaking in 1915

No account of trailblazers in the nursing profession would be complete without the inclusion of Edith Cavell, even though the greater part of her career falls outside our timescale. Also outside our timescale is her courageous and famous treatment of wounded soldiers from both sides of the conflict without discrimination during the First World War and for helping some 200 Allied soldiers escape from German-occupied Belgium and return to active service through the spy ring known as La Dame Blanche.

Edith was born in Swardeston, near Norwich, and educated at Norwich High School for Girls, then at boarding schools in Clevedon, Somerset, and Peterborough.

After time as a governess, including a spell with a family in Brussels from 1890 to 1895, Cavell came home to care for her seriously ill father which persuaded her to become a nurse. Cavell initially nursed at the Fountain Fever Hospital in Tooting from December 1895 and then applied to be and succeeded as a nurse probationer at the London Hospital in September 1896.

Cavell was seconded to work with other London Hospital nurses in the Maidstone typhoid epidemic until early January 1898, while still a probationer and was awarded the Maidstone Typhoid Medal. After her training, Cavell worked from October 1898 to December 1899 as a private nurse employed by the Private Nursing Institution of the London Hospital, treating patients in their homes. She was recommended for the position of night superintendent of St Pancras Infirmary and, in November 1903, became assistant matron of St Leonard's Infirmary in Shoreditch.

Fast forward to 1915 and Edith had been nursing troops on both sides of the increasingly bloody conflict and helping to repatriate British, French and Dutch troops to fight another day.

Her short, inspirational and courageous life ended in front of a German firing squad.

Seacole, Nightingale and Cavell were such powerful forces in the history of nursing and women's empowerment that it is easy to overlook other pioneering trailblazers who contributed much to nineteenth-century nursing. Here are two of them.

Eliza Mackenzie (1816–1892)

She was a Scottish Superintendent of Nurses appointed by the Admiralty during the Crimean War with a view to rolling out a plan to employ female nurses throughout their hospitals; we have her to thank for the formation of the Queen Alexandra's Royal Naval Nursing Service (QARNNS) in 1902.

Superintendents were not trained nurses but 'educated gentlewomen'. Mackenzie set off with her group of six nurses (Misses Vessey, Bartlett and Erskine and three others unknown) on Christmas day, 1854 for the Naval

Hospital in aptly named Therapia, near Constantinople. Of the six nurses, two were Fry nurses who had trained at the Institute for Nursing Sisters which was set up by Elizabeth Fry in Guy's Hospital in 1840. Fry nurses wore their own uniform and tended to the spiritual and physical needs of their patients.

Anne Campbell Gibson (1849–1926)

Anne Gibson was matron of the Birmingham Union Infirmary (1888–1912) and is famous for her contributions to workhouse nursing and pioneering the establishment of infirmaries separate from workhouses and staffed with trained nurses. The Poor Law Infirmary catered for over 1,000 sick, infirm and mentally ill paupers. She introduced training for nurses and put a stop to the practice of untrained workhouse inmates caring for the sick.

Pharmacists

The pharmacy profession can point to a number of other trailblazers, despite the restrictions on women in the male-dominated profession.

Rose Minshull (1845–1905)

Rose Coombes Minshull (1845–1905) was one of the first two women members, along with Isabella Clarke, of the Pharmaceutical Society of Great Britain (PSGB); she was admitted in 1879 and, as Ellen Jordan says, Rose was one of the pioneering women pharmacists who were assisted into dispensing via the Society for Promoting the Employment of Women, founded in 1859. The Society arranged for Rose, along with Louisa Stammwitz and Isabella Clarke, to be mentored by Elizabeth Garrett at her St Mary's Dispensary for Women and Children. The women were also supported by Robert Hampson (1833–1905), a champion for equality within the PSGB Council.[1]

Rose was a high-flying student, coming top of 166 candidates in the PSGB Preliminary Exam in 1873. However, Hampson's motion at the PSGB Council meeting in February 1873 to allow Minshull, Stammwitz and Alice Hart to become 'registered students' was rejected by one vote.[2] All was not lost: Minshull, Stammwitz and Isabella Clarke all studied at Dr John Muter's South London School of Pharmacy which opened in Kennington in 1870, allowing these aspiring women pharmacists to receive the comprehensive pharmaceutical education they needed and desired. Louisa Stammwitz recalled later: 'Miss Minshull and I had considerable difficulty in obtaining instruction in a chemical laboratory until Dr Muter very kindly opened his door to women.'[3]

Lab access was a real problem – and a facility jealously guarded by male pharmacists at the PSGB. Minshull, Hart and Stammwitz petitioned the PSGB Council to allow ladies access to its labs, stating in a letter to the *Pharmaceutical Journal* that 'All we ask is to be allowed the same opportunities for study, the same field for competition and the same honours, if justly won.'[4] Permission for women to work in the labs was finally granted in 1877 after Minshull and Stammwitz had passed the PSGB Minor examination, the two of them achieving the top two places in the chemistry exam. Minshull registered as a Chemist and Druggist on 18 October 1877. She passed the higher Major examination and registered as a Pharmaceutical Chemist on 19 February 1879.

After years of drawn-out debate over women's place within the PSGB, Hampson's motion to admit women to full PSGB membership was finally carried at the Council meeting on 1 October 1879, with only one vote against. Rose Minshull and Isabella Clarke were elected members of the Society.

After 1884, Rose worked at the North Eastern Hospital for Children, Hackney Road as the Dispenser. She was only too aware that women in this position were judged differently from men, writing:

> Of course she must know her work thoroughly, for she will find it to be more sharply criticised than a man's would be in the same position, keep well posted up in the current pharmaceutical literature, so as to have at her fingers' ends, when applied to, all that related to new remedies...

She summarised her long career fighting for women's recognition and just rewards as follows in *The Chemist and Druggist*:

> As the result of many years' hospital work, I am decidedly of the opinion that certainly in women's and children's hospitals a lady dispenser is the right woman in the right place.

Isabella Skinner Clarke-Keer (1842–1926)

Isabella Skinner Clarke-Keer was the first President of the National Association of Women Pharmacists. She registered with the Pharmaceutical Society as a Chemist and Druggist on 22 April 1875. Soon after, she set up her own business in Paddington, London, taking in female medical students for their dispensing course and a number of apprentices including Annie Neve (known worryingly as 'Little Arsenic') and Lucy Boole.[5] In the 1870s, Clarke was appointed as tutor in pharmacy at the London School of Medicine for Women. By 1892, new facilities at the medical school included a *materia medica* museum used 'for Mrs Clarke Keer's practical pharmacy class'.

Clarke was another high-flyer: she passed the Pharmaceutical Society of Great Britain Preliminary Examination in 1874, coming 23rd out of 355 candidates, 186 of whom failed. She gained the required three years' experience partly through time spent with Elizabeth Garrett (later Anderson) at her St Mary's Dispensary for Women and Children, and partly with Robert Hampson (1833–1905), PSGB Council member and champion of women's rights.

She passed the higher-level Major exam on 15 December 1875, the first woman to achieve this. She came fourth out of thirty-nine candidates, sixteen of whom failed.[6]

She was then appointed as Tutor in Pharmacy at the Royal Free School of Medicine for Women, but was not elected as one of the first women members of the Pharmaceutical Society until 1879.

When Alice met and married Thomas Keer, a fellow student at Muter's School of Pharmacy in Clapham, she gave up her Paddington business and joined her husband in their pharmacy in Berkeley Square.

Despite all of her and other women's achievements, it is interesting to note that the old hoary sexism prevailed: she and other women employees such as Margaret Buchanan were not allowed to work out front in the shop. Buchanan recalled that 'the presence of a woman would have been very detrimental to a pharmacist's business in those days, and more particularly to the class of business which Mr Keer conducted'. Alice suffered a 'crushing bereavement' when Thomas died, aged 46, on 14 January 1898. After his death, she began to take in boarders, later described as a Home for Students, and also ran a shorthand and typing business in Victoria Street.

Isabella later started a Home for Students at their home in Endsleigh Street, Tavistock Square, which was where the first meeting of the Association of Women Pharmacists took place.

Alice Hart (1848–1931)

Pharmacist, philanthropist, painter, textile professional and businesswoman – Alice Hart was all of these. Mary Creese tells us that Alice was educated

Mill girls in the warping room at a Wigan mill, being carefully 'overseen'.

'Debate between Suffrage & Anti-Suffrage Societies held at the Free Trade Hall, Manchester', 1909 group: not much agreement there, it appears. Sitting: Mrs Arthur Somerwell, Mrs Humphrey Ward; Margaret Ashton is third from left on the front row.

Gustave Doré was contracted to spend three months in London each year for five years to get material for his 1872 book *London: A Pilgrimage*, comprising 180 engravings with text by Blanchard Jerrold. Doré's images of the East End that 'have always drawn the most attention with their overwhelming sense of diabolic horror and epic drama, in which his figures drift like spectres coalesced from the ether'. This is 'London housing' with its crowded back-to-backs.

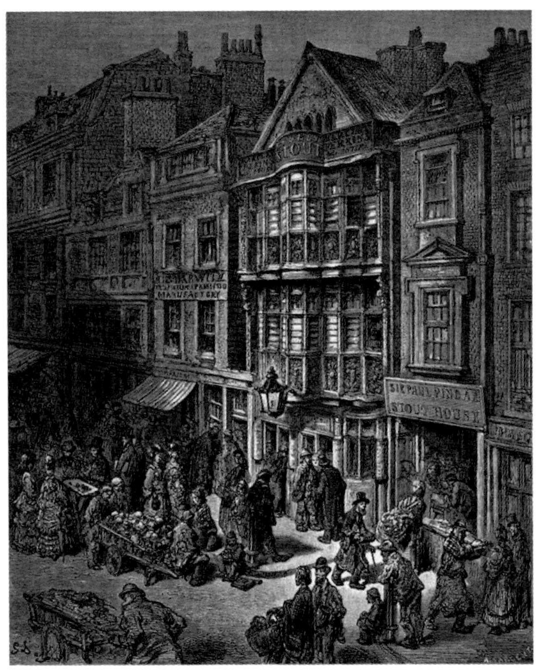

Bishopsgate with its pub and factories.

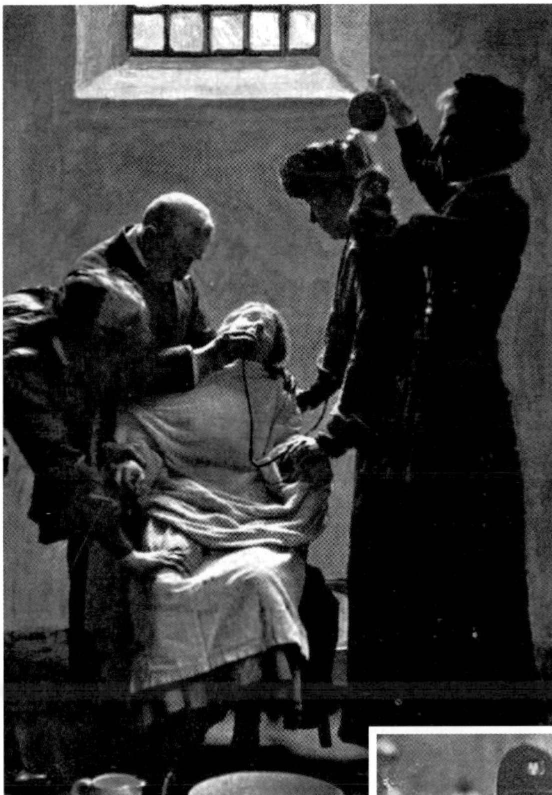

Nasal tube feeding is a way of delivering nutrition and hydration directly into a patient's stomach through a tube inserted through the nose and down the throat. It was also forced on suffragettes who went on hunger strike; Selina Martin was jailed three times for her protests. During her imprisonment in Liverpool in 1909, she was beaten, force-fed and left to freeze in her cell. 'This invasion of the body, accompanied by overpowering physical force, suffering and humiliation made many women feel they had been raped, with the words "violation" or "outrage" being commonly used.'

Dora Thewlis (1890–1976) was a suffragette whose arrest made the front page of the *Daily Mirror* and other papers.

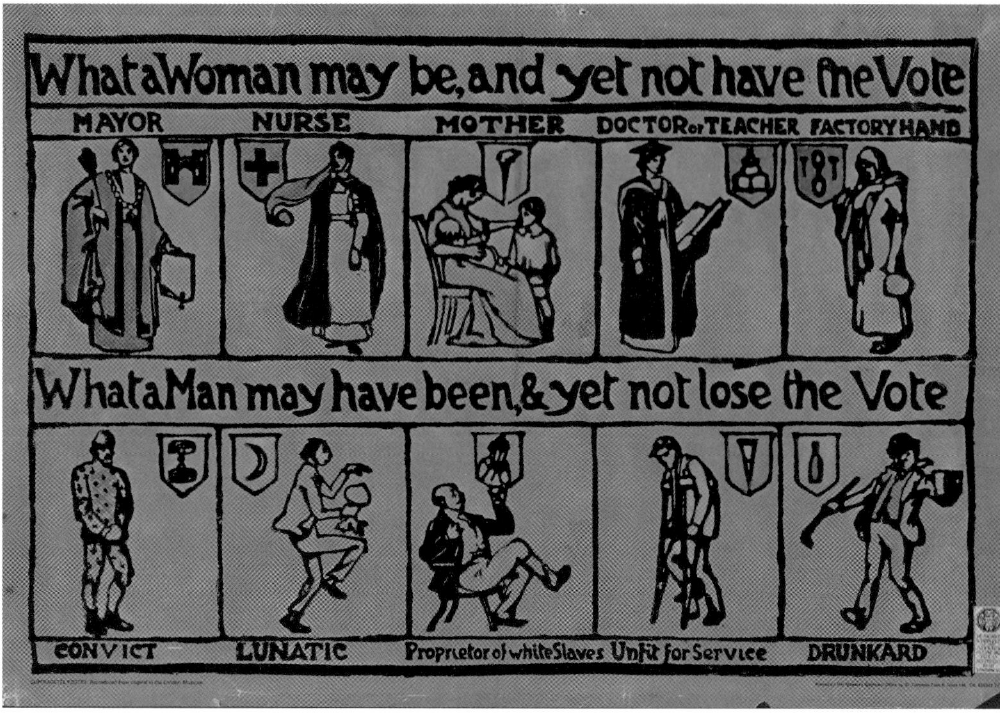

'What a woman may be and yet not have the vote'. Suffrage Atelier, an artists' collective campaigning for women's suffrage; Schlesinger Library, RIAS, Harvard University. This encapsulated perfectly the hypocrisy and double standards implicit in denying women the vote – something women had put up with for years.

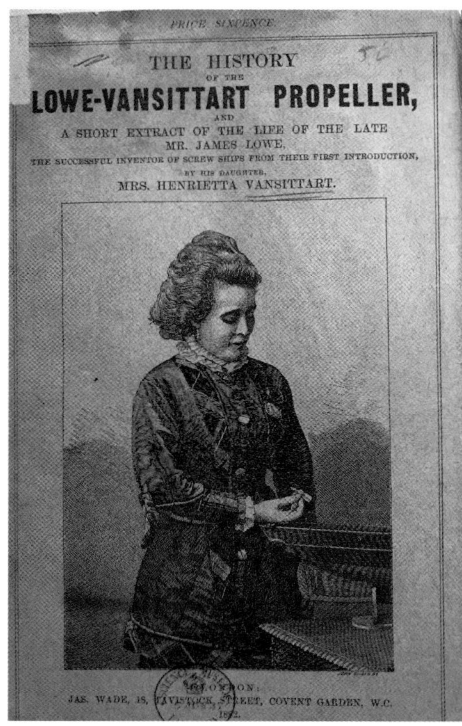

Naval engineer Henrietta Vansittart and her groundbreaking ships' propeller, *The History of the Lowe Vansittart Propeller and a short extract of the life of the late Mr James Lowe, Henrietta's father and the successful inventor of screw ships from their first introduction*, London, 1882.

Jessie Chrystal Macmillan lived a lifetime full of firsts: one of the first women to be admitted to a university course in Scotland, the first to graduate from Edinburgh in science, which she did with first-class honours in mathematics and natural philosophy; she was also the first woman to plead a case before the House of Lords, presenting her case that female university graduates should be given the right to vote. Chrystal Macmillan was also one of the founders of the Women's International League for Peace and Freedom.

The famous painting of the Peterloo Massacre, 16 August 1819, published by Richard Carlile, Manchester Libraries, Information and Archives. It shows flag waving Mary Fildes and Henry Hunt in full flow. The red plaque is on the wall of what was the Free Trade Hall, now the Radisson Hotel, in Peter Street, Manchester.

Caroline Norton, Lady Stirling-Maxwell (1808–1877) was a social reformer and author. Watercolour sketch by Emma Fergusson, 1860, National Portrait Gallery of Scotland. Norton caused a scandal when she left her husband, accused by many of coercive behaviour, in 1836. Her husband then sued her close friend Lord Melbourne, then the Whig Prime Minister, for criminal conversation (adultery). Although the jury found her friend not guilty of adultery, she failed to gain a divorce and was denied access to her three sons due to the laws at the time which favoured fathers. Norton's campaigning led to the passage of the Custody of Infants Act 1839, the Matrimonial Causes Act 1857 and the Married Women's Property Act 1870.

Kate Greenaway, 'Diamonds and Toads' published in the Frederick Warne book (1871). (*University of Florida, Baldwin Library of Historical Children's Literature*)

Milly and Olly by Mary Ward (Mrs Humphry), 1891. It is about a brother and sister who go on a mountain trip, filled with curiosity and excitement, depicting childhood innocence. (*Illustrated by Ruth Hallock*)

Anna Blunden's (1830–1915) 'The Seamstress' or 'For Only One Short Hour' (1854). Yale Center for British Art, New Haven, CT. It was the first picture Blunden exhibited publicly at the Society of British Artists in 1854; it was accompanied by a moving quotation from Thomas Hood's poem 'The Song of the Shirt' that gives voice to a 'weary and worn' seamstress as she longs for her youth in the countryside: 'For only one short hour / To feel as I used to feel, / Before I knew the woes of want, / And the walk that costs a meal!' It was influential in drawing attention to the appalling, criminal working conditions of thousands of seamstresses.

Poster advertising *Private Tinker* by John Strange Winter aka Henrietta Stannard aka Violet Whyte.

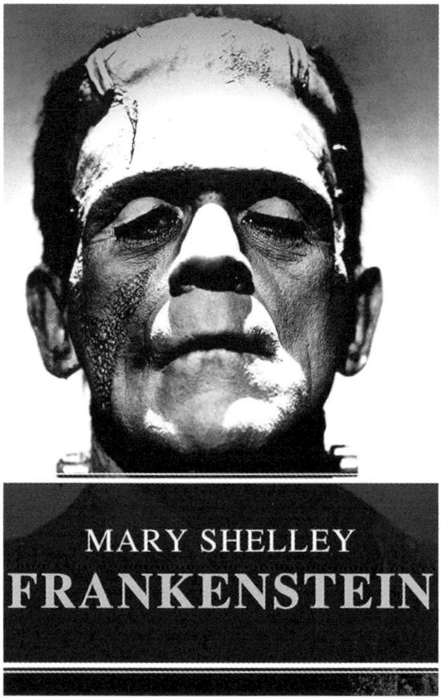

An edition of *Frankenstein; or, The Modern Prometheus* – the 1818 classic Gothic novel by Mary Shelley.

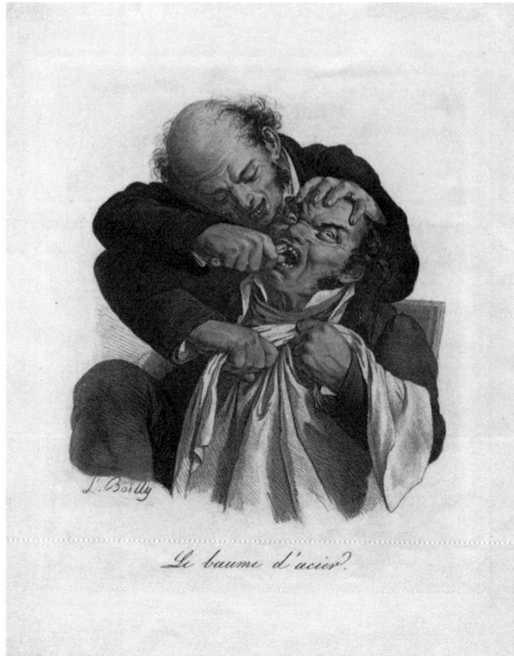

Dentistry eighteenth-century style, analgesic free. When a tooth fell out or was extracted the only way of filling that embarrassing gap was using donor teeth, usually wrenched fresh from a willing accomplice who preferred cash over his or her teeth, and implanted into the hole from which the tooth had been torn, or dentures. By Louis-Léopold Boilly, a realistic 1832 lithograph titled 'Le Baume d'Acier' (The Steel Balm) or 'The Dentist'.

Dido Elizabeth Belle Lindsay (1761–1804) and her cousin Lady Elizabeth Murray (1760–1825) at Kenwood, by David Martin. The original is in Scone Palace, Perthshire. 'The portrait of the two women is highly unusual in 18th-century British art for showing a black woman as the equal of her white companion, rather than as a servant or slave. … The basket of tropical fruit she carries and the turban with expensive feather that she wears suggest an exotic difference from her more conventionally styled white cousin, who is sitting reading a book' (English Heritage).

Capt. Frederick Forbes's portrait of Sarah as a child, from his *Dahomey and the Dahomans* (1851), 'Sarah Forbes Bonetta, the African captive', aka Aina was ward and goddaughter of Queen Victoria. She was a titled member of the Egbado clan of the Yoruba people who was orphaned during a war with the nearby Kingdom of Dahomey as a child, and was later enslaved by King Ghezo of Dahomey.

Mary Seacole and Florence Nightingale, two veritable icons of nursing who between them saved thousands of lives in the Crimean War.

South Wales mining girls; originally in Michael Hiley's *Victorian Working Women: Portraits from Life* (Boston: David R. Godine, 1980).

On the other hand, fashion was all the fashion for the middle-class fashionista. Here is the latest in garters, the Eugenie Garter, manufactured by the American Garter Company, Boston, MA and exported.

The workhouse was the destination of last resort. This extremely sad and sorry image shows people from Leeds Union Workhouse – the 'able bodied inmates'. Credit St James' Hospital, Leeds.

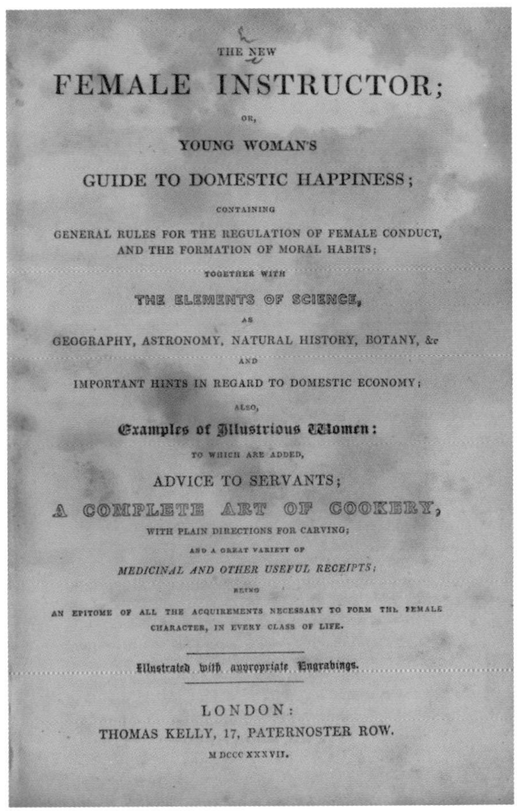

THE NEW

FEMALE INSTRUCTOR;

OR,

YOUNG WOMAN'S

GUIDE TO DOMESTIC HAPPINESS;

CONTAINING

GENERAL RULES FOR THE REGULATION OF FEMALE CONDUCT, AND THE FORMATION OF MORAL HABITS;

TOGETHER WITH

THE ELEMENTS OF SCIENCE,

AS

GEOGRAPHY, ASTRONOMY, NATURAL HISTORY, BOTANY, &c

AND

IMPORTANT HINTS IN REGARD TO DOMESTIC ECONOMY;

ALSO,

Examples of Illustrious Women:

TO WHICH ARE ADDED,

ADVICE TO SERVANTS;

A COMPLETE ART OF COOKERY,

WITH PLAIN DIRECTIONS FOR CARVING;

AND A GREAT VARIETY OF

MEDICINAL AND OTHER USEFUL RECEIPTS;

BEING

AN EPITOME OF ALL THE ACQUIREMENTS NECESSARY TO FORM THE FEMALE CHARACTER, IN EVERY CLASS OF LIFE.

Illustrated with appropriate Engravings.

LONDON:

THOMAS KELLY, 17, PATERNOSTER ROW.

MDCCCXXXVII.

For the middling and upper classes it was imperative to conduct oneself in the proper way, in all circumstances and on all occasions. Bad form was not an option. This caused anxiety and spawned a whole new genre of publishing: conduct books. Here are two examples from the thousands available.

Throughout the Industrial Revolution, those women who aspired to become medical doctors met with all manner of obstruction, sexism, discrimination, mendacity and prevarication – much of it borne of male insecurity that women might actually make as good, better even, doctors than themselves, hiding as they did behind the predictable male bastions. Here are two images which illustrate this: Pierre-André Brouillet's 'Jean-Martin Charcot demonstrating hysteria in a hypnotised patient', Salpêtière National Library of Medicine, Bethesda, New Orleans. Charcot (1825–1893) produced groundbreaking work about hypnosis and hysteria, in particular with his hysteria patient Louise Augustine Gleizes. Charcot is known as 'the founder of modern neurology'. The situation was not much better in continental Europe and in the USA: no women doctors in the audience here… Wellcome Trust Corporate Archive.

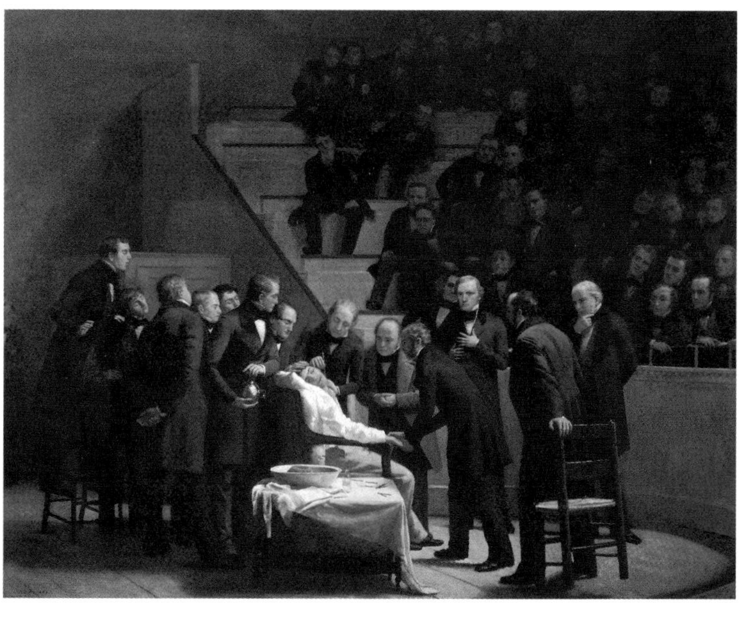

… nor here: 'First public demonstration of surgical anesthesia, by Robert Hincley', 1882, at Massachusetts General Hospital. Boston Medical Library, Cambridge, MA.

Mary Anning; doyenne of palaeontology and fossils: the Mary Anning statue in Lyme Regis, showing her holding a pointed pick hammer in her right hand and an ammonite in her left. Her dog, Tray, is in the background. www.flickr.com/photos/londonmatt/54677923862

This powerful Lydia Becker banner was designed by Mary Lowndes, leading stained-glass artist who was an active participant in the suffragette movement, acting as Chair of the Artists' Suffrage League, and creating poster art to assist the movement.

Mary Ellen Best. Kitchen sink social history artist *par excellence*. Here we see her *Artist at Work* and the still life *The Sideboard*. She is famous for painting scenes of domestic life, depicting families at tables, in kitchens, and domestic staff in her various homes in York, Nuremberg, Frankfurt, Worms, and Darmstadt. *The Artist at Work* and *The Sideboard* are in York Art Gallery; courtesy of York Museums Trust.

Floriography, the language of flowers (and love) was all the rage amongst the better off in the Industrial Revolution. This colour lithograph *Langage des Fleurs* from the *Album de la Décoration* by Alphonse Mucha (1898) captures the essence of this crypto courtship fad with its abundant lilies.

'Miss Garrett Anderson before the board of Medical Examiners at Paris' where she eventually received her MD in 1870. National Library of Medicine, Bethesda, New Orleans; wood engraving published in *Harper's Weekly*, 23 July 1870. Anderson and others went to Paris to enable them to qualify as doctors as it was impossible in England or Scotland at the time.

An example of Anna Maria Garthwaite's sumptuous and vivid floral designs for silk fabrics handwoven in Spitalfields. https://nataniabarron.com/2021/09/14/anna-maria-garthwaite-spitalfields-silk

A 'man-midwife' represented by a figure divided in half, one half representing a man and the other a woman. 'An important address to wives and mothers on the dangers and immorality of man-midwifery'. New York Academy of Medicine. Coloured etching by I. Cruikshank, 1793. Wellcome Collection.

Mary Delany. In 1771, by now a widow in her early 70s, Mary took up decoupage, all the rage with fashion ladies of the court. The results were detailed and botanically accurate depictions of plants on tissue paper coloured by hand. She created 985 of these beautiful works, her 'Paper Mosaiks [*sic*]', from the age of 71 to 88, when her eyesight failed her. http://alchetron.com/Mary-Delany-1083539-W

at the Apothecaries Society in London and went on to study medicine in Paris.[7] One of her sisters, Fanny, was badly brain-damaged. Another sister was Dame Henrietta Rowland Barnett who also spent time in philanthropy. Together the two sisters taught the poor at Toynbee Hall.

An 1883 trip to County Donegal with her husband opened Alice's eyes to the crushing poverty there. She decided that the best way to have an impact was to revive the local cottage industries, especially weaving tweed. By the end of 1883, the Donegal Industrial Fund had been founded and the next year Donegal tweed was shown at the London International Health Exhibition and went down so well that Hart opened a shop at 31 New Cavendish St, London. Hart worked closely with the local weavers, sending them designs and examples of Scottish tweeds and by experimenting with dyes from local plants which won a Sanitary Institute of Great Britain medal.

In 1885, Hart took it all to a higher level, getting Irish women to set up embroidery schools which became the 'Kells embroidery' schools all over the country, teaching the embroidery of flax on linen, making use of designs from Irish manuscripts and Japanese art.

The items produced at the Kells schools were displayed at the London International Inventions Exhibition in 1885 and earned them a gold medal. By 1886, the Donegal Industrial Fund had moved to bigger premises at 43 Wigmore St, London. A grant allowed for further schools to teach weaving and embroidery. In 1888, at the Irish Exhibition, she set up 'Donegal industrial village'. Farm labourers from Donegal were brought over to train at what was then Regent Street Polytechnic.

Throughout all of this, Hart painted. After a trip to Burma (Myanmar) with her husband, she wrote a successful book on the people and the country.

Notes

1. Ellen Jordan (June 2001), '"Admitting ... a Dozen Women into the Society": the first women members of the British Pharmaceutical Society', *Pharmaceutical Historian*, 31(2), pp. 18–27; Ellen Jordan (1998), '"The great principle of English fair-play": male champions, the English women's movement and the admission of women to the Pharmaceutical Society in 1879', *Women's History Review*, 7(3), pp. 381–410.
2. Justin Lincoln (25 September 1926), 'The Late Mrs Clarke-Keer', *The Chemist and Druggist*, 105, p. 530.
3. 'Lady-pharmacists', *The Chemist and Druggist*, 145, 30 July 1892.
4. S. W. F. Holloway, *Royal Pharmaceutical Society of Great Britain 1841–1991: A Political and Social History*, Pharmaceutical Press, p. 263, 1991.

5. Lucy Everest Boole FRIC was a British chemist and pharmacist who was the first woman to research pharmacy in England. She was the first female professor at the London School of Medicine for Women in the Royal Free Hospital, and the first female Fellow of the Royal Institute of Chemistry.
6. *The School of Pharmacy, University of London: Medicines, Science and Society, 1842–2012*, Elsevier Science, p. 65, 2014.
7. Mary R. S. Creese, *Ladies in the Laboratory? American and British Women in Science, 1800–1900: A Survey of Their Contributions to Research*, Scarecrow Press, pp. 151ff, 200.

Dentists

And there were women dentists too. We have Jane Austen to thank for a myriad of things in Regency society which started to benefit us through the eighteenth and nineteenth centuries and in the years of the Industrial Revolution. Dentistry is definitely not one that immediately springs to mind, but the years under discussion here saw the rise of dentistry as a science and a profession, and as a palpable benefit to the aching public, opening wide in their thousands. Women played a major part in this oral revolution.

Unlike frocks, hair styles, modern languages, keyboard skills and table manners, teeth did not feature in the catalogue of mandatory accomplishments for young ladies; in fact, it was the done thing to keep your mouth shut whenever possible in order not to display the horror shows decaying behind the lips of many a young, and not so young, lady. Flashing a smile was poor form.

But that all started to change in 1787, when artist Madame Vigée-Lebrun caused a scandal by unveiling a smiling self-portrait in the Paris Salon… in which she revealed a set of beautifully white teeth. This was a shock of the first order: one disgusted contemporary condemned it as 'an affectation … which shows no precedent amongst the Ancients'.[1]

Jane Austen obviously thought dentistry to be worthy of comment in her world; a dental appointment probably meant a trip to London or some other metropolis with the prospect of no anaesthetic and lots of clumsy ironwork. We know of such a nightmare trip when Austen accompanied three of her nieces to a Mr Spence, London dentist, and how it inspired a similar trip where the victim was Harriet Smith, in *Emma*.[2]

> The poor girls and their teeth! I have not mentioned them yet, but we were a whole hour at Spence's, and Lizzy's were filed and lamented over again, and poor Marianne had two taken out after all, the two just beyond the eye teeth,

> to make room for those in front. When her doom was fixed, Fanny, Lizzy, and I walked into the next room, where we heard each of the two sharp and hasty screams.
>
> The little girls' teeth I can suppose in a critical state, but I think he must be a lover of teeth and money and mischief, to parade about Fanny's. I would not have had him look at mine for a shilling a tooth and double it. It was a disagreeable hour.
>
> —Jane Austen to Cassandra, 16 September 1813

For the overwhelming majority of people, of course, dentistry was a rudimentary do-it-yourself affair with a little amateur help from friends and family, and the occasional farrier or blacksmith. This form of domestic torture has, sadly for many, enjoyed a revival in some parts of England.

Oral hygiene then was in its infancy: toothpicks may have been fashionable, but toothbrushes were thin on the ground – usually frayed sticks with hog's hair; whiteners took the form of toast or toxic sounding-charcoal or similar such abrasives. The dentists' armamentarium comprised instruments of torture left over from the Middle Ages: first-line treatment was extraction by pelicans and keys – the former was a brutal instrument with a pad or bolster, which was placed on the side of the gum below the tooth to be extracted and a beak or claw which engaged the opposite side. A downward twist of the handle tore the tooth out of the mouth. The key was similar, but had a handle similar to that of a corkscrew and enabled the instrument to be used more comfortably from the front of the mouth instead of the side. The 1860s saw these tools replaced by forceps.

The only way of filling the embarrassing gap was donor teeth, usually wrenched fresh from a willing accomplice who preferred cash over his or her teeth, and implanted into the hole from which the tooth had been torn, or dentures. Before 1850 dentures generally did not fit well.

Our saviour was Pierre Fauchard (1678–1761), who authored the must-have, go-to bible of dental practice, the 1723 *Le Chirurgien dentiste; ou, traite des dents* (*The Surgeon-Dentist; or Treatise on Teeth*). He was the first to practise dentistry scientifically, and to explicate practice management techniques to increase patient comfort, and advocate what today we call oral or dental health or hygiene. Still no effective pain control though.

Here are some of the women private dental practitioners who were busy taking some of the pain out of well-to-do living.[3]

In 1738, Ann De La Mare, the widow of James De La Mare, 'operator for teeth to his Royal Highness the Prince of Wales', 'had gone into partnership with a Mr John Baptist Landies, the son of Mr Landies, operator for teeth, in Paris', who 'draws, cleans and sets artificial teeth etc in the best manner'.

In 1777, a Mrs Levis and her husband, both '*surgeon dentists in all its particular branches*', were running their business from Queen Street, Bath. Mrs Levis would attend the ladies and Mr Levis, the gentlemen. Free advice on procedures would be given for all difficult cases.

Also in 1777, a Mrs De St Raymond was working from her home, No. 9, Kings-square Court, Soho. A bit of an all-rounder, she was recommending her many services to the nobility and gentry:

> Her well known skill in the performance of chirurgical operations, for the various disorders of the mouth; especially the lightness of her hand, in removing all tartarous concretions, destructive to the teeth, and her dexterity in extracting stumps, splints and fangs of teeth. She also draws, fastens, fills up and preserves teeth, corrects their deformity, transplants the fore-teeth from one mouth to another. Likewise grafts on and sets in human teeth; makes and fixes in artificial teeth, from one to an entire set, and executes her newly invented masks for the teeth, and obturators for the loss of the palate.

Catherine Madden of 53 St John's Street, West Smithfield was working as a dentist between 1790 and 1799, whose cures were apparently so efficacious that she guaranteed '*no recurrence of the trouble*'.

In 1792, there was a Mrs Hunter, who worked from her home, London's 78 Great Titchfield Street. Not only was she a dentist, but she also 'treated people's complexion', so a side hustle beautician too. She claimed to be able to relieve tooth ache and prevent it from returning without the need for extractions. She especially commended her services to women, who may prefer to be attended to by another woman. She also specialised in treating children as she had 'a gentle touch', which would make the process less terrifying.

She charged one guinea at the start of treatment and then four guineas per annum, which would include tooth powder and tincture; or half a guinea for each consultation after the first, and half price for children.

Notes

1. Quoted in *The Regency Dentist*, 2011, The Jane Austen Centre: contactus@janeausten.co.uk.
2. Ibid. And https://janeaustensworld.com/2022/10/05/dentistry-during-jane-austens-era.
3. All cited by Sarah Murden, 'Female dentists of the 18th century', 26 September 2019, in *Georgian Medicine & Health: Women's History*.

Entrepreneurs and Inventors

The laws on intellectual property and patents during the Industrial Revolution were another barrier to the innovative and entrepreneurial woman, given that a man inherited any patents held by the woman, be she wife, daughter or next of kin. In 200 years, *The Times* published five articles on the subject. In fact, many earlier references to 'the woman inventor' in English newspapers were actually about Americans. The first patent awarded to an Englishwoman was Amy Everard Ball's in 1637 (patent no. 104,) just seventy-six years after Elizabeth I had awarded the first patent, which is held at the British Library. Ball claimed to possess 'the mystery, skill, and invention' of making essential tinctures from saffron, roses, and other flowers. As a widow, she was allowed to pay the patent fees in instalments.

In 1709, Mrs Susanna Centlivre, author of *The Busie Body* Restoration comedy wrote 'Want [is] the mistress of invention,' as the many women who strived to improve their lives by designing more comfortable clothing would have agreed.[1] The patent records are replete with new and improved corsets and cycling ware that didn't affront modesty so much so that the British Patent Office started to publish statistics for this category at the end of the nineteenth century.

Note

1. *The Busie Body* was first performed at the Drury Lane Theatre in 1709. Margo Collins tells how it focuses on the legalities of what constitutes a marriage, and how children might subvert parental power over whom they can marry. *The Busie Body* has the distinction of being the most popular female authored-play of the eighteenth century, and became a stock piece of most theatres during the period.

Jane Vanef

The second patent was awarded to Jane Vanef of Westminster in 1737 for her collapsible, hinged hoop: 'A way of compressing the hoops of skirts to allow for entry and exit into carriages, churches and other enclosed spaces.' Women of the upper classes owe Vanef a great deal: the voluminous skirts they habitually wore on their perambulations or in cramped carriages on the way to the next ball could now be tamed and that unwieldy 5 foot of material kept under a modicum of control.

Mrs Clements (fl.1720) – cutting the mustard

The years of the Industrial Revolution were witness to many an innovation in food: Alfred Bird invented baking powder (and later, custard powder); Cadbury's, York's Rowntrees and Terry's, Caley's of Norwich and Fry's perfected drinking chocolate and then chocolate confectionery while Sheffield's Henry Henderson was producing his eponymous Henderson's Relish and in 1814 Jeremiah Colman founded Colman's of Norwich to manufacture and market his famous hot English mustard.

But much earlier than that, a little known 'Mrs Clements' had invented dry mustard: in 1720 in Saddler's Yard in Durham City, she discovered that grinding mustard seeds made it taste a lot stronger. Word spread and she introduced her new recipe to King George I, who liked it so much that he patented the product and made it fashionable. As every good sales rep should, Mrs Clements, ever keen as mustard, travelled the length and breadth of England taking orders for her hot product. Cruikshank satirised it in his 'Durham mustard too powerfull for Italian capers, or the opera in an uproar'. The business later was bought by Colmans of Norwich; when Colman's commissioned a new trademark in 1855, they acknowledged Mrs Clements by choosing the head of the famous Durham ox.

Eleanor Coade (1733–1821)

A truly trailblazing businesswoman famous for her manufacture of countless Neoclassical statues, architectural decorations and garden ornaments made of *Lithodipyra* (Coade stone) from 1769 until her death.

Lithodipyra ('stone fired twice') was a top-quality, durable, moulded weather-resistant stoneware; statues and decorative features made from this still look almost new today. Her material resisted the corrosion from

London's (and other cities') polluted air, contributed to by exhausts and chimney smoke and the formation of acid rain.

Here are some of her major achievements: she produced stoneware for

- St George's Chapel, Windsor
- The Royal Pavilion, Brighton
- Carlton House, London
- The Nelson pediment at the Royal Naval College, Greenwich
- The frontispiece of the original Twinings tea shop on the Strand
- Private ornaments in the rear of Buckingham Palace
- The lion on Westminster Bridge
- The statue of George III on the seafront at Weymouth, Dorset
- The stone awning and statues at Schomberg House, Pall Mall
- The crest on the Imperial War Museum
- Shortly after her death, her company produced much of the stoneware used in the refurbishment of Buckingham Palace

She conducted her business, 'Coade's Artificial Stone Manufactory', 'Coade and Sealy' and later 'Coade' by appointment to George III and the Prince Regent, for half a century in Lambeth.

In her will, Coade left most of her fortune to charity schools and clergymen, as well as to her family. A champion of women's rights, she also left money to a handful of married women friends, insisting that their husbands were to have no control over the money.

This is how Nicola Phillips summed up Coade's impressive and rare legacy:

> Coade's success as a businesswoman was very rare in the Georgian era. She was a hard-working individual who concentrated on methodical procedures to produce consistently high quality products. She was the first and only person to succeed in the artificial stone business, thanks to a combination of managerial skills, entrepreneurial flair, and a talent for marketing and public relations.

Dorothy Thomas (1756–1846)

Dorothy Thomas (also Dolly Kirwan or Doll Thomas), former Caribbean slave known as 'the Queen of Demerara', is certainly one of the most driven, entrepreneurial and active women in a book full of driven, entrepreneurial and active women. Her life and achievements are full of paradox and surprise as this short summary will show.

The Centre for the Study of the Legacies of British Slavery at UCLA puts it well:

> She deftly navigated a difficult world to obtain freedom for both herself and her large family. The success she was able to achieve in not just one colony, but several, is extraordinary. Dorothy Thomas sits therefore at the apex of a list of free coloured and free black female entrepreneurs who were able to exploit the fluid situation in the Atlantic world during the age of revolutions and achieve remarkable success. The education she so assiduously provided for her children and grandchildren enabled them to exploit their advantaged position and find a niche in the heart of the British empire.

Before she arrived at Dominica, she had given birth to three children. She was illiterate,[1] but soon became known for her business acumen; she prudently insisted that her children be recorded with their correct fathers' surnames in their records for manumission.[2]

Dorothy Kirwan was born as a slave between 1756 and 1763 in Montserrat to Betty, a slave owned by the Kirwan family. Dorothy's owner was Andrew Kirwan. Because she was given the family surname, raised as a Catholic, and is described as mulatto, despite being of black complexion, we can assume she was fathered by a member of the Kirwan family.[3]

The first thing Dorothy did as soon as she could was to buy her freedom. Where did the money come from? Around 1781, she was purchased from Andrew Kirwan by William Foden, who fathered two of her children, William and Charlotte Kirwan.[4] Kirwan arranged for the manumission of her son Edward on 24 July 1781. Kirwan paid for the two transactions and it seems she may have earned the money by having 'the freedom to sell her person as a courtesan to the planter class'.[5] Kirwan also purchased the freedom of her two eldest daughters.

Foden died in 1782 and in 1784, under the provisions of the will, she was left his household goods and one-third of his livestock. Additionally, she was discharged from any debts she might owe him and the will specified that she had paid to him money to manumit herself and her three children, William, Charlotte, and a baby, Ann.

She then conducted business in Montserrat, Dominica, Grenada, Barbados, and Demerara and spent nearly sixteen years securing the freedom of her children, mother, and other relatives. She went on to own hotels, one of which boasted a French restaurant, but her main income stream was hiring out female hucksters to whom she supplied goods to be sold on to plantation workers and slaves. What's more, she hired out her slaves as labourers, earned income from lodging houses, ran a plantation, and rented out other properties which she owned. Dorothy was one of the few black women who benefited from the British government scheme to reimburse slave owners: she received £3,413 for the loss of her slaves when Britain abolished slavery.

Thomas was a frequent visitor to London where she ensured that many of her descendants received a British education and that her daughters were all married off to prominent white businessmen; she was as ever on hand to bail out the families when they were in financial straits. An avid networker and socialite, she enjoyed influence among a wide circle of business and influential connections in the colonies of the British West Indies despite her illiteracy and used her networks to good effect when required. She left both a large and prominent family scattered throughout the British Empire, and a significant historical legacy which shines a crucial light on free women of colour and their lives in her era.

Dorothy Thomas was nothing short of a legend and many an apocryphal story was written about her, including accounts of visits to both houses of Parliament and a personal audience with George IV. Candlin concludes that

> She lived in an era when war opened opportunities to free women of colour, but threw them back into obscurity when it ended. Thomas' life is illustrative of the difficult and complex society that existed in the British Caribbean. It offers significant insight into the system of slavery and manumissions, as well as the difficulties and success she faced as a free woman of colour and a business woman. The fact that she was able to be successful in multiple colonies shows her ability to use the fluidity of changing times to her advantage.[6]

Notes

1. Kit Candlin, 'Thomas, Dorothy "Dolly" (c. 1763–1847), prominent free colored businesswoman and slave owner'. In Franklin W. Knight and Henry Louis Gates Jr. (eds.), *Dictionary of Caribbean and Afro–Latin American Biography*, Oxford, 2016a.

2. Kit Candlin and Cassandra Pybus, 'Enterprising Women and War Profiteers: Race, Gender and Power in the Revolutionary Caribbean'. In Alan Forrest, Karen Hagemann and Michael Rowe (eds.), *War, Demobilization and Memory: The Legacy of War in the Era of Atlantic Revolutions*, Basingstoke, 2016, pp. 106–107.
3. Candlin and Pybus, 2016, p. 106; Candlin, 2016a.
4. Candlin and Pybus, 2016, pp. 105–108.
5. Candlin and Pybus, 2016, p. 107; Kit Candlin, 'Thomas [formerly Kirwan], Dorothy'. *Oxford Dictionary of National Biography* (online ed.), Oxford, 2016b.
6. Candlin, 2016a.

Henrietta Caroline Bentley (fl.1794–1820)

Inventor Henrietta Caroline Bentley had, in the years before 1820, been granted a pension of £300 per year (about £21,000 today). In 1794, she, a spinster of Southampton and no doubt every nurse's and patient's champion, was granted a patent for a 'bed for invalids, which may be made and the linen changed without inconveniencing the patient'.

Elizabeth Bell (fl.1803–1807)

More social and industrial historians should know about Elizabeth Bell, a nineteenth-century inventor who did her best to ameliorate the disgusting working conditions inflicted on climbing boys and girls in the eighteenth and nineteenth centuries. It would be hard to beat the cruelty these boys and girls had to endure in the name of profit in a supposedly civilised country (even then).[1] Before we talk about Bell's invention, here is what the child sweeps might have to endure on a daily basis: a vivid description of the fate of a climbing boy illuminating the terrifying darkness, the claustrophobia and 'accidental' death by asphyxiation:

> After passing through the chimney and descending to the second angle of the fireplace the Boy finds it completely filled with soot, which he has dislodged from the sides of the upright part. He endeavours to get through, and succeeds in doing so, after much struggling as far as his shoulders; but finding that the soot is compressed hard all around him, by his exertions, that he can recede no farther; he then endeavours to move forward, but his attempts in this respect are quite abortive; for the covering of the horizontal part of the Flue being stone, the sharp angle of which bears hard on his shoulders, and the back

> part of his head prevents him from moving in the least either one way or the other. His face, already covered with a climbing cap, and being pressed hard in the soot beneath him, stops his breath. In this dreadful condition he strives violently to extricate himself, but his strength fails him; he cries and groans, and in a few minutes he is suffocated. An alarm is then given, a brick-layer is sent for, an aperture is perforated in the Flue, and the boy is extracted, but found lifeless. In a short time an inquest is held, and a Coroner's Jury returns a verdict of 'Accidental Death'.[2]

Chimney sweeps' cancer, described as a 'canker of privities', is a squamous cell carcinoma of the skin of the scrotum. It has the dubious distinction of being the first reported form of occupational cancer, initially identified by Percivall Pott (1714–1788) in 1775, who established that the cancer was triggered by occupational exposure to soot, hence its prevalence amongst chimney sweeps. His work helped lead to the Chimney Sweepers Act 1788 and his cutting-edge *Chirurgical Observations* helped shape the modern understanding of occupational cancers.

Lord Shaftesbury led the campaign in Parliament. Pott's work provoked a series of 'Chimney Sweepers' Acts' which paved the way for the formation of societies such as the 'Society for Superseding the Necessity of Climbing Boys, by Encouraging a New Method of Sweeping Chimneys and for Improving the Condition of Children and Others Employed by Chimney Sweepers' in 1803, which enjoyed royal and aristocratic patronage. Despite this, however, and public demands to transform chimney sweeping practices, for example suggesting that mechanical devices take the place of climbing boys, insurance companies and homeowners argued that chimney sweeps' carcinoma was a small price to pay for the protection of the population at large against chimney fires. In spite of the call for action as early as 1788, Pott's work only began to influence change in 1875, when a decline in scrotal cancer incidence and mortality was noted.

Cancer apart, there was a multitude of life threatening, life-shortening chronic maladies and injuries to contend with in later life: falls from great heights causing deformed ankles, broken legs, and twisted spines. Eye and respiratory problems were frequent due to blinking through and breathing noxious soot and fumes.

Climbing boys, then, were just so much collateral damage. In 1824, the Society for Superseding the Necessity of Climbing Boys, along with the Society for the Encouragement of Arts, Manufactures, and Commerce (which survives today as the Royal Society of Arts) offered a prize for any machinery that would end the 'pernicious employment' of climbing boys. George Smart's mechanical 'chimney sweep' machine from 1805 differed little from the brushes already used by the climbing boys. However, he went on to win the prize offered by the Society with his jointed articulated brush, called a 'scandiscope'. Smart's 'scandiscope' was more flexible and portable, and could be operated from the comfort and safety of the hearth. Many rubbished Smart's device as impractical, however, and so climbing boys continued to climb, suffer terrible injuries and psychological trauma, and, in some cases, die.

Notes

1. In H. A. Waldron, 'A brief history of scrotal cancer', *British Journal of Industrial Medicine*, 40(4), 1983, pp. 390–401 (p. 391).
2. The heinous trade in child slavery relating to sweeps is dealt with in horrific detail in Chapter 7: Climbing Boys: Up the Chimney, in Paul Chrystal, *Factory Girls and Climbing Boys: Women and Children at Work from the Industrial Revolution to 1900.*

Margaret Knight (1838–1914) – the world's first bag lady

If you've ever wondered who invented the ubiquitous paper bag then wonder no more: it was Margaret Eloise Knight, who did everyone everywhere an inestimable favour with this world- beating item of convenience for almost anything, from groceries to, well, anything. The paper bag was there for us again in the twenty-first century when the insidious everlasting land-filling plastic bag went out of favour.

Often called 'the most famous nineteenth-century woman inventor', Margaret founded the Eastern Paper Bag Company in 1870; she acquired thirty patents in different fields and has become a symbol of women's empowerment.

Sadly, when Knight patented her paper bag machine, women held a tiny fraction of patents; 150 years later, nothing much has changed: in 2019 less than 10 per cent of primary inventor patent holders were female.

Hertha Ayrton (1854–1923)

Some describe Hertha Ayrton (née Phoebe Sarah Marks) as 'Britain's foremost women inventor', others 'the most distinguished woman scientist' in the country at that time. From the age of 9, she was educated by her aunt Marion Hartog, who along with her husband ran a school in London where they educated family children. Hertha studied French and music, the standard subjects of a prospective governess, but she added mathematics and Latin taught to her by talented male cousins. At 16, Hertha started as a live-in governess in London, sending money home to her family in Portsea. Hertha knew that she could do better. And she did do better.

The Jewish Women's Archive tells us that Ottilie Blind

> encouraged Hertha to study for the Cambridge University Local Examinations for secondary schools, which had been opened to girls in 1865 after a campaign by Emily Davies and Barbara Bodichon. Hertha passed the examination in 1874, taking honours in English and mathematics.

She could count on no personal wealth but was able to attract the attention and financial support of well-positioned patrons, including George Eliot, which enabled her to study mathematics at Girton College, Cambridge.

Ottilie took Hertha to suffrage meetings where she met Barbara Bodichon (1827–1891), the leader of the Langham Place feminist activists. In 1869, Bodichon and Davies had co-founded the first university college for women, Girton College, Cambridge. After interviewing Hertha, Bodichon paid for advanced mathematics lessons and arranged a loan so that she could attend Girton.[1] Bodichon supported her throughout her education and career, including bequeathing her estate to Ayrton.[2] Ayrton had celebrated Bodichon by naming her first child, a daughter born in 1886, Barbara Bodichon Ayrton-Gould (1886–1950). The daughter was called 'Barbie', and she later became Labour MP for Hendon North from 1945 to 1950 after giving up her career in science teaching to join the WSPU in 1906.

What else did Hertha Ayrton do while at Cambridge?

- She constructed a sphygmomanometer (blood pressure monitor).
- She led the choral society.

- She established the Girton fire brigade.
- Together with Charlotte Scott, she formed a mathematical club.

What degree was Hertha Ayrton awarded? In 1880, she passed the Mathematical Tripos, but Cambridge did not grant her an academic degree because, as we know only too well, at the time it did not award full degrees to women and would not until 1948. She was awarded a Third Class degree, the ignominious fate of all successful women at the time at Cambridge and for long after. As a married woman, she had no legal existence in arcane and hoary British law. Ayrton then passed an external examination at the far more enlightened University of London, which awarded her a Bachelor of Science degree in 1881.[3]

Hertha Ayrton was brought up as Jewish but was agnostic by her teens. She adopted the name 'Hertha', a nickname given by Ottilie Blind, after the eponymous heroine of a poem by Swinburne that criticised organised religion. Others say Ottilie nicknamed Sarah 'Hertha' after the ambitious heroine of a novel by the feminist Swedish writer, Frederika Bremer.[4]

On graduating, Hertha kept herself busy: she earned pin money by teaching and embroidery, ran a club for working girls, and looked after the needs of her invalid sister. She taught maths at Notting Hill and Ealing High School, and devised and solved mathematical problems, many of which were published in 'Mathematical Questions and Their Solutions' from the *Educational Times*.

She studied electrical engineering at evening classes at Finsbury Technical College given by Professor William Edward Ayrton, a pioneering inventor in electricity, whom she married. She published influential papers in *The Electrician* and in 1899 was the first woman ever to read a paper before the Institution of Electrical Engineers (IEE): 'The Hissing of the Electric Arc'. Soon after, Ayrton was elected the first female member of the IEE; the world had to wait until 1958 for Dorothy Smith, the next woman to be admitted to the IEE. Hertha's nomination for membership at the Royal Society was rejected because married women were ineligible, but she was nevertheless awarded a prestigious prize, the Hughes Medal, by the Royal Society 'for her work on electric arcs and ripple marks in sand and water'.

Suffragists paid for Hertha Ayrton's 1884 British and US patents for 'an elegant mathematical line divider consisting of a series of parallelograms, designed for the purpose of dividing a line into any number of equal parts' and for enlarging and reducing figures; its primary use was for artists for enlarging and diminishing, but it also proved useful to architects and engineers. Barbara Bodichon and Lady Goldsmid (another supporter of the 1866 suffrage movement and a benefactor of significance to Girton) together advanced her enough to take out those patents. This divider was shown at the Exhibition of Women's Industries and attracted much media attention.

Later she saved countless soldiers' lives during World War I when she invented a simple hand-held device for dispersing poison gas in the trenches. Towards the end of the war, about 100,000 fans were sent to the front as a result of her own persistent lobbying. She later applied similar principles for the propulsion of noxious gases in industry.

Her most significant technological advances owed much to her electricity research, including anti-aircraft searchlights and arc lamps. Ayrton's 1884 patent was the first of many – from 1884 until her death, Hertha registered twenty-six patents: five on mathematical dividers, thirteen on arc lamps and electrodes, the rest on air propulsion.

Another important string to Hertha's bow was her relentless support of women's suffrage. It all started when she was in her teens, joining the WSPU in 1907, but her work with Emily and Christabel Pankhurst and her friendship with Marie Curie and her daughter Irène, to whom she gave maths tuition, sadly falls out of the scope of this book. Suffice to add that Hertha was outraged when Marie's discovery of radium was attributed to her husband: 'she conducted a vigorous campaign in the press, commenting that "errors are notoriously hard to kill, but an error that ascribes to a man what was actually the work of a woman has more lives than a cat."'[5]

Notes

1. Nina Byers and Gary A. Williams (eds.), *Out of the Shadows: Contributions of Twentieth-Century Women to Physics*, Cambridge, 2006.
2. 'Hertha Marks Ayrton', *cwp.library.ucla.edu.*
3. Larry Riddle, 'Hertha Marks Ayrton', *Biographies of Women Mathematicians*, Atlanta, GA: Agnes Scott College, 2016.
4. https://jwa.org/encyclopedia/article/ayrton-hertha-marks
5. https://jwa.org/encyclopedia/article/ayrton-hertha-marks

Journalists

Anna Blackwell (1816–1900) – 'Fidelitas'

Born in Bristol, Blackwell was a writer, journalist and translator specialising in spiritual and social issues; she enjoyed a long (forty-two years) successful career as Paris correspondent of leading papers like *Once a Week*, *English Woman's Journal* and *The Ladies' Repository*. Blackwell also wrote poetry, fairy tales, and essays on occult subjects. As a teacher and journalist, she had a wide influence in the US and in France.

Of her sisters, Elizabeth was the first woman to receive a medical degree in the United States, while Emily was the third. Sarah was an artist. Their sister-in-law, Antoinette Brown Blackwell, was the first woman to be ordained as a Protestant minister in the United States.

As a spiritualist in 1873, the *Eclectic Magazine* announced that Blackwell had printed a pamphlet entitled 'Spiritualism and Spiritism', featuring 'some rather strange revelations'. Blackwell had informed the editors that she had authentic evidence, revealed to her by two spirits, that so far back as the year 3543 BCE she had been nothing less than Princess of Abyssinia.

Lucie Heaton Armstrong (1851–1907)

Lucie Armstrong was an Anglo-Irish journalist and writer, chiefly on etiquette.

Her first work, *The Ball-Room Guide*, was published anonymously in 1880; soon afterwards she had stories in Cassell's *Little Folks* magazine for children. Some of the these were reprinted in her second book, *Doll Stories*, which came out in 1883. She published in several journals, magazines and newspapers, often anonymously or under pseudonyms, and published nine books, eight of which were on etiquette. The etiquette market at the time was huge, catering as it did for the dizzying world of romance, balls and social visiting in which knowing how to conduct oneself was *de rigueur*.

In 1893, Armstrong published *The Etiquette of Party Giving*, in which she not only provided the etiquette of giving a party, but also went further, delineating several types of parties, including 'A Clover Tea', 'A Cobweb Party', 'A Palette Party' and 'An Epithet Party'.

Armstrong died at the Camberwell House Asylum.

These were her published books: *The Ball-Room Guide* (1880) (updated in 1900); *Doll Stories* (1883; as Lucie Cobbe); *The Etiquette of Party Giving, etc.* (1893); *Good Form. A Book of Every Day Etiquette* (1889); *Etiquette for Girls* (1893); *The Etiquette of Party Giving with Hints to Hostess and Guest* (1893); *Letters to a Bride, Including Letters to a Débutante* (1896); *Etiquette-up-to-Date* (1898); and *Etiquette and Entertaining* (1903).

Catherine Drew (1832–1910)

Another Anglo-Irish writer and journalist, Catherine Drew grew up in Belfast and Dublin, beginning her journalist career writing for the *Irish Builder* and eventually becoming its assistant editor. She went on to write for the *Belfast Newsletter*, and moved to London in 1871 as the paper's London correspondent where she wrote two columns, 'Metropolitan gossip' and 'Ladies' letter'. These were among some of the earliest regular columns written specifically for women, providing society news for her readers in Belfast. Her articles were also published in *The Literary World*, *The British Architect* and *London Society*.

Drew was one of the founding members of the Ladies' Press Association, campaigning for greater rights for women journalists, and was a leading light in the Institute of Journalists, representing the Institute at several international congresses.

In 1894, she was one of the signatories of the Frances Power Cobbe memorial, campaigning for greater recognition and rights for women journalists, alongside Millicent Fawcett and Jessie Boucherett. Drew published a number of novels, including *Harry Chalgraves's Legacy* (1876) and *The Lutanistes of St Jacobi's* (1881). In 1885, she gave a lecture titled *Dress, economic and technic* at the Exhibition of Women's Industries in Bristol, which was later repurposed as a pamphlet.

Politicians

Mary Fildes (1789–1876)

People wonder when they look at contemporary images of the Peterloo atrocity who that brave and committed woman was up there on the stage passionately waving her banner and urging on the crowd. Standing with the main act, speaker Henry 'Orator' Hunt, 27-year-old Mary Fildes is up there dressed in Sunday-best white, gripping her banner pole, and sporting a red Cap of Liberty, with her banner depicting Justice triumphant trampling Corruption underfoot: how ironic when all the trampling done on that day was done by Corruption, killing and injuring those who were there in the name of Justice.

Fildes was president of the Manchester Female Reform Union and mother of five children. Her prominent position on the platform and her exhortations would make her a target of violence from the authorities.

Katherine Connelly, in her article 'The Women of Peterloo', explains how the special constables, who arrived to arrest the speakers, beat her with their truncheons. When she tried to escape from the platform, her dress caught on a nail and, whilst trapped and suspended above the ground, she was slashed with a sabre by a soldier who seized her banner as a trophy.

Here is the text from a 'very fragile newspaper cutting (in 3 pieces) from The *Manchester Guardian*, Tuesday, June 30, 1908'.

> To Henry Hunt, Esq., as chairman of the meeting assembled in St. Peter's Field, Manchester, sixteenth day of August, 1819, and to the female Reformers of Manchester and the adjacent towns who were exposed to and suffered from the wanton and fiendish attack made on them by that brutal armed force, the Manchester and Cheshire Yeomanry Cavalry, this plate is dedicated by their fellow labourer, Richard Carlile.

The original coloured engraving vividly depicts the Peterloo Massacre and all its brutality. All the poles from which banners are flying have Phrygian caps or liberty caps on top. However, not all the details strictly accord with contemporary descriptions; the banner the woman is holding should read: 'Female Reformers of Roynton – Let us die like men and not be sold like slaves'.

Mary Fildes's exposed position on the platform, and the way she was targeted by the authorities, testified to the vital role that women played in this struggle for democracy.[1]

Notes

1. www.counterfire.org/article/the-women-of-peterloo/#:~:text=The%20woman's%20name%20was%20Mary,beat%20her%20with%20their%20truncheons

Frances Wright (1795–1852)

Widely known as Fanny Wright, she was a Dundee-born lecturer, writer, free thinker, feminist, utopian socialist, abolitionist, social reformer, and Epicurean philosopher, who became a US citizen in 1825. Before that, however, after her parents had died, a maternal aunt became Wright's guardian and taught her ideas relating to the French materialists. In 1813, when Wright was 16, she returned to Scotland to live with her great-uncle, James Mylne, a philosophy professor at Glasgow College. Wright took an interest in the Greek philosophers, especially Epicurus, the subject of her first book, *A Few Days in Athens* (1822), which she had written by the age of 18. Wright also studied history and became interested in the United States' democratic form of government.

Her visits to the USA in 1819 and 1824 nurtured her support for universal education, the emancipation of slaves, birth control, equal rights, sexual freedom, legal rights for married women, and liberal divorce laws; she also vocalised her opposition to organised religion and capital punishment.

Mary Clifford (1842–1919)

Clifford was a trailblazer for women serving on Boards of Guardians and developed a widely popular scheme for the fostering of orphans. She was a founder member of the National Union of Women Workers (NUWW), and served as its president from 1903 to 1905.

She firmly believed that many orphans would fare better if they were relocated to Canada, and developed an emigration scheme with Mark Whitwill. She was a force behind reductions in the power of abusive parents over their children, which were passed into law in 1889. She was co-opted to the Central Committee of Poor Law Conferences, serving on it for twelve years.

Jane Cobden Unwin (1851–1947)

The daughter of Richard Cobden and younger sister of suffragette Anne Cobden-Sanderson, Jane grew up in a household where radical politics were ingrained and the legacy of her father's work was passionately kept alive by Unwin and her sisters. A founding member of the Women's Suffrage League, she also championed Irish home rule, the abolition of slavery, campaigned for the families of strikers in East London and for the rights of indigenous people in colonial territories. She was one of the first women controversially elected into the new London county council in 1889, sitting for Bow and Bromley.

Both parents left their girls in no doubt regarding their responsibilities for the poor in the local community. In her 1864 diary, Jane Cobden records visits to homes and workhouses. She and Anne, at the ages of 12 and 10 respectively, taught in the local village school. The girls were encouraged by their father to contribute what money they had to relieve local poverty: 'Do not keep the money … as you have now made up your minds to give it to poor sufferers, let your own neighbours have it. Your Mama will tell you how to dispose of it, and tell me all about it.'[1]

Notes

1. Sarah Richardson, 'You Know Your Father's Heart: The Cobden Sisterhood and the Legacy of Richard Cobden'. In: Howe, Anthony and Morgan, Simon (eds.), *Nineteenth Century Liberalism: Richard Cobden Bicentenary Essays*, Aldershot, 2006, pp. 235–236.

E. Nesbit (1858–1924)

We remember Edith Nesbit as a prolific author of forty popular books for children, the most famous of which was *The Railway Children*. But E. Nesbit also composed poetry (her first love) and adult supernatural fiction (for example *Man-size in Marble and Other Grim Tales*) and was a co-founder of the Fabian Society with her husband Hubert Bland. She was also an active lecturer and prolific writer on socialism during the 1880s and also wrote with her husband, with whom she conducted a free marriage, under the name Fabian Bland.

Other supernatural books are *From the Dead: The Complete Weird Stories of E. Nesbit*, 2005; and *The Head*.

Her political writings include: 'Women and Socialism: from the Middle-Class Point of View', *Justice*, 4 and 11 April 1885, and 'Women and Socialism: A Working Woman's Point of View', *Justice*, 25 April 1885.

Nesbit has the father in *The Railway Children* declare that 'girls are just as clever as boys, and don't you forget it!' but she did not support women's rights: 'She opposed the cause of women's suffrage – mainly, she claimed, because women could swing Tory, thus harming the Socialist cause.'[1]

Notes

1. Jessica Winter (28 September 2022), 'The British writer who rewrote the world for children', *The New Yorker*.

Alys Mallt (1867–1950)

Alice Matilda Langland Williams, more commonly known as Alys Mallt and Y Fonesig Mallt, was a fervent Welsh political writer and celtophile. She was born in Brecknockshire to an English-speaking family, but learned Welsh when influenced by the circle of Welsh heiress Augusta Hall, Baroness Llanover.

Throughout her adult life, Mallt was involved in the ongoing campaign for Welsh home rule. She routinely supported Welshness in dress and speech and also joined Welsh nationalist organisations, such as Cymru Fydd and Plaid Cymru. She admired the enthusiasm of Irish nationalists over the water. Her brother Frederick dressed as a thirteenth-century Welsh lord.

In 1889, when she was 22, Williams co-published her first novel with her sister, Gwenfreida, under the pseudonym 'Y Ddau Wynne'. The novel, titled *One of the Royal Celts*, is an idealised fictionalisation of the British Empire, in which Celtic bravery gets the credit for the success of the empire. There are no English characters in the book.

Her 1901 *A Maid of Cymru* is a highly patriotic romance allegorised with the failure of the campaign for Welsh home rule as shown by the death of the *Tangwysh* who was trampled to death by wild ponies.

Women's Suffrage

Readers who have made it this far will be aware that many of the women trailblazers explored so far entertained leanings towards the women's suffrage movement to some degree and in one form or another. By the end of the Industrial Revolution, signed-up suffragettes and suffragists were many and numerous amongst the educated middle class in particular and membership of both the Women's Social and Political Union (WSPU) and the National Union of Women's Suffrage Societies (NUWSS) was growing apace. Its rise and rise is explored here in the section on 'Women and the Vote' above.

Here is a small selection of trailblazing women born before 1864 who worked effortlessly and courageously to ensure women won an equal place in society with the rights they deserved.

Lydia Becker (1827–1890)

We know of the Pankhursts' work in fighting for women's suffrage but there is another Manchester woman who deserves much credit in championing that cause, and that woman is Lydia Ernestine Becker, who as well as being a leader in the early British suffrage movement was a passionate amateur scientist with interests in biology and astronomy.

In 1867, she founded the Ladies' Literary Society in Manchester. She corresponded with Charles Darwin and soon afterwards convinced him to submit a paper to the society. Becker sent plant samples to Darwin taken from the fields around Manchester and gave Darwin a copy of her 'little book', *Botany for Novices* (1864). Becker is one of a number of nineteenth-century women who contributed to Darwin's scientific work.

Botany remained important to her, but her work for women's suffrage took over: her involvement in promoting and encouraging scientific education for girls and women brought these two things together.

She it was who established Manchester as a centre for the suffrage movement when she convened the first meeting of women's suffrage; and with Richard Pankhurst she got the first woman to vote in a British election: a court case was unsuccessfully brought to exploit the precedent. The first public meeting of the National Society for Women's Suffrage was held on 14 April 1868, in the Free Trade Hall, at which Becker moved that women should have voting rights on the same terms as men. Becker also published the *Women's Suffrage Journal* between 1870 and 1890 with Jessie Boucherett.

Lydia was opposed to the violent and media friendly means espoused by Emmeline Pankhurst, which explains why she has been virtually forgotten in women's history.

Arguing there was no difference between the intellect of men and women, Becker was a vocal advocate of a non-gendered education system in Britain. She also differed with many suffrage activists in that she argued for the voting rights of *unmarried* women as well as women in general.

The Lydia Becker Institute of Immunology and Inflammation was opened at the University of Manchester in 2018 – home to internationally renowned immunology and inflammation expertise and research.

Her works included:

- *Botany for Novices* (1864)
- 'Female Suffrage' in *The Contemporary Review* (1867)
- 'Is there any Specific Distinction between Male and Female Intellect?' in *Englishwoman's Review of Social and Industrial Questions* (1868)
- 'On the Study of Science by Women' in *The Contemporary Review* (1869)
- 'The Political Disabilities of Women' in *The Westminster Review* (1872)

Barbara Bodichon (1827–1891)

Independent income gave Bodichon a freedom rare for most women and she and like-minded ladies met regularly in the 1850s to discuss women's rights – 'The Ladies of Langham Place' – one of the first organised women's movements in Britain. In 1854, she published *Brief Summary of the Laws of England concerning Women*, which facilitated the passage of the Married Women's Property Act 1882.

In 1858, Bodichon established the *English Women's Journal*, which urged manual or intellectual employment and equality for women, expansion of employment opportunities, and reform of laws relating to women.

In 1866, along with Emily Davies, Bodichon initiated a scheme to extend university education to women which developed into Girton College, Cambridge, to which Bodichon gave freely of her time and money.

Painting was a passion; she studied under William Holman Hunt. Her watercolours were exhibited at the Salon and the Royal Academy and were admired by Corot and Daubigny. She was an early member of the Society of Female Artists (SFA), showing fifty-nine artworks with them between 1858 and 1886. She was a close friend of George Eliot's.

Josephine Butler (1828–1906)

If any of our trailblazers give us a rude shock when they remind us that many of the social, political and legal challenges facing women and children in the late nineteenth century are still plaguing us today, it is Josephine Butler. She campaigned for women's vote, better education for women, the end of coverture in English law, the repeal of the Contagious Diseases Acts, the abolition of child prostitution and stopping human trafficking of young women and children into European prostitution.

When the last of her four children (Eva) tragically died falling from a banister in August 1864, Butler refocused her life to helping others, starting with the inmates of a local workhouse and then campaigning for women's rights in British law. Eva fell 40 feet from the top-floor banister onto the stone floor of the hall; she died three hours later.

Having moved to Liverpool, Butler later wrote that she 'became possessed with an irresistible urge to go forth and find some pain keener than my own, to meet with people more unhappy than myself. … It was not difficult to find misery in Liverpool.' She made regular visits to the workhouse at Brownlow Hill, which could accommodate 5,000 persons. Here she would sit down with the women in the cellars – many of whom were prisoners – and join them picking oakum, while discussing the Bible or praying with them.

Butler ardently campaigned for women's rights: their right to vote and a better education. In 1866 she was a signatory on a petition to widen the franchise to include women. The petition, which was supported by the MP and philosopher John Stuart Mill, went ignored.

In 1867, with the suffragist Anne Clough, she established the North of England Council for Promoting the Higher Education of Women, which aimed to professionalise the status of governesses and female teachers; she served as its president until 1873. The next year she published *The Education and Employment of Women* in which she argued again for the provision of higher education for women, and more equal access to a wider range of jobs. It was the first of her ninety books and pamphlets. That May she petitioned the senate of the University of Cambridge to offer examinations for women and the Cambridge Higher Examination for women came in the following year. Jordan notes that 'much of the credit for this should go to Anne Clough, but … Butler played a very influential part … of the campaign.'[1]

In 1869, she became heavily involved in the campaign to repeal the Contagious Diseases Acts, legislation that clumsily attempted to control the spread of sexually transmitted infections – especially in the British Army and Royal Navy – through the non-consensual internal examination of alleged prostitutes, a highly humiliating and deeply discriminatory procedure she described as 'surgical or steel rape'. No evidence was required, other than the say-so of the police officer. If a magistrate agreed, women were given those undignified digital genital examinations. If a woman was found to have an STI, she was detained in a locked hospital until the condition was cured. If they refused to be examined or hospitalised they could be imprisoned, often with hard labour thrown in. In relation to her opposition to the Contagious Diseases Acts, Butler wrote:

> You must know there are many good men and women in our country who have devoted their lives to the work of reclaiming prostitutes, and of offering protection and aid to women and young girls, who through poverty, ignorance, or evil companionship are in danger of falling into sin. And because several persons working together can do more than each working alone, societies have been formed for this purpose, one of which, the Rescue Society, has in the last seventeen years, opened the doors of its various Homes to no less than 6,722 fallen women and girls, of which number seventy out of every hundred have been restored to a virtuous life, whilst lack of funds has compelled it reluctantly to refuse admission to many others who implored its aid.[2]

Units of plain-clothed policemen swarmed around certain areas, specialising in arresting suspected prostitutes; according to Jordan, the officers were 'hated for their surveillance and harassment of prostitutes and working-class women ... who they treated with little regard for their legal rights'.[3] Women who were subjected to the examination found their names and reputations tainted and, according to the historian Hilary Cashman, 'the Acts had the effect of turning them to prostitution by barring respectable ways of life to them'.[4] Getting a job after a conviction would have been nigh on impossible, with one of the very few ways of earning money left open to these women being... prostitution. The Acts were repealed in 1886.

In the course of her research, Butler was horrified to find that some of the prostitutes were as young as 12, and that there was a thriving slave trade in prostitution. A campaign to halt the trafficking led to the dismissal of the chief of the Belgian *Police des Mœurs*, and the trial and imprisonment of his deputy and twelve brothel owners, all implicated in the sordid trade.

Butler fought child prostitution with the help of the editor of *The Pall Mall Gazette*, William Thomas Stead, who, incognito, bought a 13-year-old girl with the collusion of her mother for £5. The subsequent outcry led to the Criminal Law Amendment Act 1885, which raised the age of consent from 13 to 16 and brought in a raft of measures to stop children descending to prostitution. Her final campaign was in the late 1890s, against the Contagious Diseases Acts, which continued to be implemented in the British Raj, a law unto itself in many respects.

Butler helped form the Ladies National Association for the Repeal of the Contagious Diseases Acts (LNA) and published a *Ladies Manifesto*, which highlighted the discriminatory nature of the legislation:

> [it] not only deprived poor women of their constitutional rights and forced them to submit to a degrading internal examination, but they officially sanctioned a double standard of sexual morality, which justified male sexual access to a class of 'fallen' women and penalised women for engaging in the same vice as men.[5]

In her determined pursuit of the repeal, Butler had toured Britain in 1870, covering 3,700 miles to attend ninety-nine meetings during the year. Her target audience was working-class family men, most of whom were outraged at the description Butler gave of the examination women were forced to endure: that 'surgical or steel rape' mentioned above. She faced major opposition and danger. At one meeting, pimps pelted her with cow dung; at another, the windows of her hotel were smashed, while at a third, threats were made to raise the building where she was speaking.

Notes

1. Jane Jordan, *Josephine Butler*, London, 2001, p. 96.
2. In Sheila Jeffreys (ed.), *The Sexuality Debates*, London, 2001.
3. Jordan, 2001, p. 104.
4. Hilary Cashman, 'Singular Iniquities: Josephine Butler and Marietta Higgs', *New Blackfriars*, 71(834), 1990, pp. 26–32 (p. 27).
5. Judith R. Walkowitz, 'Male Vice and Feminist Virtue: Feminism and the Politics of Prostitution in Nineteenth-Century Britain', *History Workshop*, 13(13), 1982b, pp. 79–93 (p. 80).

Elizabeth Wolstenholme-Elmy (1833–1918)

Elizabeth was especially keen to expose marital rape for what it was: a legal inequality and a form of sex slavery.

Wolstenholme-Elmy spent the best part of her life as a significant campaigner and fixer in the history of women's suffrage. She published essays and some poetry, using the pseudonyms 'E' and 'Ignota'.

She was at the heart of almost every Victorian feminist campaign, ranging from the demand for better education, the right to vote and the rights of prostitutes to the sensitive issue of marital rape.[1]

Maureen Wright identifies a life-changing moment in Elizabeth's young life:

> But Elizabeth defied her guardian and studied privately, preparing herself to be a governess and, latterly, headmistress of her own girls' school [The Grange]. She had no desire to remain in the domestic realm. She placed her commitment to feminism from the moment when, acting as a bridesmaid aged 17, she fully realised what marriage meant for women – a 'lifelong sentence of pauperism and dependence' with no control over their actions or autonomy over their own bodies.

Wright makes the significant point that regarding her work at The Grange, it is clear that the curriculum she taught was not one only of female 'accomplishments' (such as singing, dancing and drawing) but included political economy, mathematics and other skills thought to be too 'masculine' in nature for a girls' school. Elizabeth sought to equip her girls for not only the world of marriage and motherhood, but for the world of work, and many of them did in fact go on to become headmistresses of schools.

Amongst numerous other affiliations, she was a Member of the Manchester National Society for Women's Suffrage, the Manchester Branch of the Society for the Employment of Women and the Northern Counties League for the Repeal of the Contagious Diseases Acts.

It was, however, Elizabeth's secularism that caused the greatest challenges for many of her contemporaries: she was an advocate of 'free love', a concept which became particularly problematic when she fell pregnant. Living together while unwed obviously subverted the staid moral values of mid-Victorian society.

We have seen how Elizabeth was especially keen to expose marital rape for what it was: a legal inequality and a form of sex slavery. Why did she feel so strongly about this issue? Maureen Wright explains:

> Elizabeth's abhorrence of marital rape became clear in 1880, when she stood on the platform of the London Dialectical Society to declare her desire to see the practice criminalised. Her opposition was in part built on personal reasons and a desire to see a legal inequality quashed.

Wives were often beaten or starved for non-compliance or, as evidence from one notable legal case of 1891 shows, imprisoned against their will. Elizabeth saw the crime of marital rape as one common to women of all classes, and thus a cause of unity. At a moment when even polite society was concerned with the relentless rise in sexually transmitted diseases, she found a receptive audience, in some quarters, for her views. Of all the disadvantages married women faced, Elizabeth believed this 'sex slavery', as she termed it, to be the worst.

For all her efforts, Elizabeth did not see a law passed against it in her lifetime – in fact this did not reach the statute book for another 100 years, in 1991.[2]

Notes

1. Sarah Irving, 'Elizabeth Wolstenholme-Elmy: Manchester's Free Love Advocate and Secular Feminist' (8 April 2012), *Manchester's Radical History* (wordpress.com).
2. Marital rape finally became a criminal offence in 1991 following the landmark House of Lords ruling in the case of R v. R, which overturned the long-standing legal precedent that a husband could not be convicted of raping his wife, citing a marital rape exemption based on the idea of implied consent through marriage.

Dame Millicent Garrett Fawcett (1847–1929)

> 'A wise constant and courageous Englishwoman. She won citizenship for women.'

In fact, she campaigned all her adult life for women's suffrage by legal means and in 1897–1919 was leader of Britain's largest women's rights association, the National Union of Women's Suffrage Societies (NUWSS). Her explanation: 'I cannot say I became a suffragist. I always was one, from the time I was old enough to think at all about the principles of Representative Government.' She did inestimable good work when she tried to broaden women's opportunities for higher education, as a governor of Bedford College, London (now Royal Holloway) and co-founding Newnham College, Cambridge in 1871. In 2018, 100 years after the Representation of the People Act, she was the first woman to be honoured with a statue in Parliament Square.

Ray Strachey tells us 'The Garretts were a close and happy family in which children were encouraged to be physically active, read widely, speak their minds, and share in the political interests of their father, a convert from Conservatism to Gladstonian Liberalism, a combative man, and a keen patriot.'[1]

How Millicent Garret embarked on her particular road in life is quite disarming in its matter-of-factness and pragmatism. When still a child, Fawcett's elder sister Elizabeth Garrett Anderson, Britain's first female doctor, introduced her to Emily Davies, an English suffragist. In her mother's biography, Louisa Garrett Anderson quotes Davies as saying to her mother, to Elizabeth and to Fawcett: 'It is quite clear what has to be done. I must devote myself to securing higher education, while you open the medical profession to women. After these things are done, we must see about getting the vote.' She then turned to Millicent: 'You are younger than we are, Millie, so you must attend to that.'[2]

And that's exactly what Millie spent the rest of her life doing.

John Stuart Mill introduced her to Henry Fawcett, a Liberal Member of Parliament who had designs on marrying her sister Elizabeth until she decided to concentrate on her medical career. Millicent and Henry married in 1867. Henry had lost his sight in a shooting accident in 1858. Their marriage was said to be based on 'perfect intellectual sympathy'; Millicent pursuing her writing career while caring for Henry, and running two households: in Cambridge and London. The couple were strong believers in proportional representation, free trade, and advancement for women. Their only child, Philippa Fawcett (b. 1868), became the first woman to get the top score in the Cambridge Mathematical Tripos exams.

In 1868, Fawcett joined the London Suffrage Committee, and in 1869 spoke at the first public pro-suffrage meeting held in London. Her *Political Economy for Beginners* was 'wildly successful'. Fawcett was a founder member of the National Vigilance Association, a moral purity campaign brought to prominence by Stead's 1885 exposure of the white slave trade in London. 'Stead Hostels' were named in his honour to house and protect young working women and girls.[3]

Notes

1. Ray Strachey, *The Cause: A Short History of the Women's Movement in Great Britain*, CreateSpace Independent Publishing Platform, 2016.
2. Louisa Garrett Anderson, *Elizabeth Garrett Anderson, 1836–1917*, London, 1939.
3. T. Little, 'Millicent Garrett Fawcett, 1847–1929', *Journal of Liberal History*, Liberal Democratic History Group, 2020; L. Delap, '"No one pretends he was faultless": W. T. Stead and the Women's Movement', Open Library of Humanities, 2013.

Margaret Ashton (1856–1937)

Born in Withington, Manchester, Margaret Ashton was an English non-militant suffragist, local politician, pacifist and philanthropist, and the first woman city councillor for Manchester on her election as councillor for Manchester Withington. She worked relentlessly for the peace movement nationally and internationally and was a well-known speaker on public platforms across the country. Later, she was a signatory to the Open Letter, a call for peace addressed in sisterhood 'To the Women of Germany and Austria'. She initiated a Manchester branch of the Women's International League for Peace and Freedom.

Ashton supported mother and baby clinics and urged the provision of free milk for babies and new mothers.

Lily Maxwell (1800–1876)

> 'Lily is more than a mistake on the electoral roll. This Scottish former servant is part of the fabric of the long, hard fight for women to get the vote.'
>
> —Sophy Ridge

You could be forgiven for believing that the Pankhurst family was largely responsible for one of the greatest changes in British electoral history. And you'd be right, but Lily Maxwell, a humble widowed shop owner, runs them a close second when she appeared in error on the register of voters in Manchester. She was originally in domestic service before saving up enough money and setting up a shop selling a wide range of goods from crockery and candles to red herring at 25 Ludlow Street, Chorlton-upon-Medlock. Issues relating to electoral change, however, was not the first time she had brushed with the law of the land. On 4 April 1866, Lily made the newspapers when she was fined £1 in the Police Court for defrauding her customers with falsified weights and light measures.

Because the properties in Ludlow Street were expensive enough to qualify their occupiers under the pre-1867 £10 household borough franchise, Lily's name appeared on the list of voters for Manchester – her name was mistaken for that of a man. Her accidental inclusion on the list was spotted by one of the candidates for election, Jacob Bright, an ardent supporter of women's suffrage. Along with his wife, Ursula Mellor Bright, he was a founder of the Manchester branch of the National Society for Women's Suffrage, in January 1867.

Bright's election committee alerted Lydia Becker, the society's secretary, to Maxwell's appearance on the list, and she informed Maxwell of the situation. The outcome was a pivotal test case at the Court of Common Pleas. Becker visited Maxwell and went with her to the polling station in Chorlton Town Hall. The returning officer found Maxwell's name on the list and allowed her to vote; reputedly the room erupted with cheers for Britain's first ever female voter.

In January 1868, the *Englishwoman's Review* wrote of Lily:

> It is sometimes said that women, especially those of the working class, have no political opinion at all. Yet this woman, who by chance was furnished with a vote, professed strong opinions and was delighted to have a chance of expressing them.

Having left her mark on the nation's electoral history and women's suffrage, Lily was largely forgotten. She died in 1876 in the Withington Workhouse.

For more on Lydia Becker, see above.

Rosalind Howard (1845–1921), Countess of Carlisle

Remembered as 'The Radical Countess', she was an inveterate campaigner for women's suffrage and temperance reform. Rosalind took the temperance pledge in 1881 and closed down all public houses on her estates, including Castle Howard and Naworth Castle, the following year. Castle Howard became the venue for annual temperance demonstrations and galas, and in 1903 she became the President of the National British Women's Temperance Association (NBWTA). After twelve months' campaigning for the temperance movement, over 1,500 people had signed the temperance pledge. She pioneered the Scientific Temperance Instruction, which encouraged schools to teach children about the dangers of alcohol.

As well as her valuable temperance work and raising eleven children, she helped managed the family estates, and was an active member of local government. In 1890, she became a member of the Women's Liberal Federation and persuaded the organisation to support extending the suffrage to all women.

Dora Montefiore (1851–1933)

In 1889, when her husband was lost at sea, Montefiore learned that she had no automatic right to guardianship of her children; it was then that she became an advocate of women's rights, serving on the executive of

Millicent Fawcett's National Union of Women's Suffrage Societies and joining the Women's Social and Political Union as formed by Emmeline and Christabel Pankhurst.

In 1897, Montefiore floated the formation of the Women's Tax Resistance League. In 1906, to protest against the lack of political representation, Montefiore declined to pay her taxes and remained stubbornly barricaded in her home for six weeks – wonderful PR for the League.[1]

The house, surrounded by a wall, could be reached only through an arched doorway, which Montefiore and her maid barred against the bailiffs. For six weeks, Montefiore resisted payment of her taxes, addressing the frequent crowds through the upper windows of the house.[2]

A banner was displayed across her wall proclaiming: 'Women should vote for the laws they obey and the taxes they pay.'[3] She said, 'I was doing this because the mass of non-qualified women could not demonstrate in the same way, and I was to that extent their spokeswoman. It was the crude fact of women's political disability that had to be forced on an ignorant and indifferent public.'[4]

That same October, Adela Pankhurst, Montefiore and others were arrested for demanding votes for women in the lobby of the House of Commons. Montefiore later described the experience and the interior of Holloway Prison.[5] Montefiore also joined several socialist organisations around this time, including the Women's Freedom League, the Social Democratic Federation and the British Socialist Party. Montefiore was also a notable linguist and published the first English translation of the works of Maxim Gorky.

Notes

1. David M. Gross, *99 Tactics of Successful Tax Resistance Campaigns*, Picket Line Press, 2014, pp. 44, 94, 189.
2. Laura E. Nym Mayhall, *The Militant Suffrage Movement: Citizenship and Resistance in Britain, 1860–1930*.
3. Hannah Awcock (19 July 2018), 'Turbulent Londoners: Dora Montefiore, 1851–1933', *Turbulent London*.
4. John Simkin, 'Dora Montefiore', Spartacus Educational.
5. 'From a Victorian to a Modern', www.marxists.org.

Isabella Ford (1855–1924)

'Freedom is what women cry for, not philanthropy.'

Isabella Ormston Ford was a social reformer, suffragist and writer on socialism, feminism and workers' rights.[1] She was born in Adel Grange in Adel, Yorkshire, the youngest of eight children in a pacifist Quaker family. Her father was a solicitor who ran a local night school for mill girls. Contact with these inspirational girls gave Ford and her sisters an insight into class differences and an interest in their parlous working conditions. When she was 16, she began teaching at her father's school.

Ford was mainly active in Leeds, where she worked tirelessly amongst tailoresses who were campaigning for improved working conditions; she helped Emma Paterson, president of the WPPL, to form a Machinists' Society for tailoresses in Leeds – the opening shot by Ford to improve the pay and conditions of women in the West Yorkshire textile industry. Isabella was a member of the Fabian Society, and in 1897 she worked with the Leeds Socialist League, assisting in industrial disputes. In 1889 she was instrumental in setting up the Leeds Tailoresses' Union, becoming its President and addressing large meetings as well as being active in its strike.

Isabella was a founder member of the second Leeds Women's Suffrage Society, and also joined the Women's Emancipation Union, which encouraged women to stand for election to local administrative bodies: she became a parish councillor for Adel-cum-Eccup.

In 1890, Isabel marched alongside workers from Manningham Mills (aka Lister's Mill) which in its heyday was employing 11,000 men, women and children – all busy turning out top-end textiles made from quality fabrics in velvet and silk. In 1911, the firm supplied 1,000 yards of velvet for George V's coronation.

Lister's was involved in the formation of the Independent Labour Party (ILP) with the 1890–91 strike at the mill. When US import tariffs began to impact on the bottom line, Lister's simply and capriciously used the expedient of posting a notice in the mill announcing cuts in the wages of 1,100 workers by 25 per cent in the run-up to Christmas, with the threat of locking out those who refused. The Victorian equivalent of inhuman resources by email. The badly organised but determined workforce were furious and at its height the consequent strike saw 5,000 workers downing tools from December to April with riots in the streets midst futile attempts to negotiate a settlement.

Lister, meanwhile, equivocated and dissimulated; he even went so far as to threaten to close down the mill completely and move production to Addingham near Ilkley; he then denounced his workforce in a letter to the local press, exposing his ignorant prejudices based on stereotypical views of the working classes and their alleged innate fecklessness: 'The women spend their money on dress and the men on drink, so that the begging box goes round – it matters not what wages are.' His verbal arrogance turned out to be a highly effective recruiting sergeant for the trade unions and for the nascent Independent Labour Party.

The striking workers were denied any help from the Poor Relief so it was not long before poverty and hunger drove the strikers back to work in April 1891, giving Lister a complete victory and preserving his precious profit margins.

During the 1890s, Isabella was active in propaganda work for the ILP throughout the West Riding of Yorkshire, speaking 'often at street corners, in dingy club rooms, in hot, crowded school rooms' (*Bradford Pioneer*, July 1924). She was the author of numerous pamphlets including *Women's Wages* (1893), *Industrial Women and How to Help Them* (1900) and *Women and Socialism* (1904).

As well as being a founder member of the Independent Labour Party, she was a published novelist, and oversaw the Leeds Women's Suffrage Society becoming part of the new National Union of Women's Suffrage Societies in 1898.

Ford campaigned for animal rights and was a vegetarian. She was an anti-vivisectionist and was Chair of Leeds RSPCA. In 1896, Ford added her signature to the Humanitarian League's petition against vivisection.

In 1902, Isabella organised the Yorkshire and Cheshire Women Textile Workers suffrage petition; 33,184 female textile workers in Yorkshire signed, and she travelled to London with the delegation to present it to Parliament.

Ford was a perceptive commentator on contemporary events unfolding around her: here are some of her observations.

On women and trade unions:

> Some unions excluded women, and others admitted them, mostly in order to prevent them from being blacklegs to men. 'The women must be got in because they [men] undersell us, they injure us' – it was the old story of Rousseau and the sole object of women's education being to make women useful to men, over again ... Women resent this spirit of antagonism between them and men.
>
> —*Why I Joined the Independent Labour Party* (1896)

On sexual harassment at work; not much has changed here:

> A premium is sometimes put on impropriety of conduct on the women's part by the foreman. That is, a woman who will submit or respond to his coarse jokes and language and evil behaviour receives more work than the woman who feels and shows herself insulted by such conduct, and wishes to preserve her self respect. The pittance earned by some of these women is earned at the expense of more than only hard toil. Even when this coarseness is confined to language only, it causes deep suffering to some of the women. They feel, they know, that because they are women and therefore regarded as helpless and inferior, they are spoken to as men are not spoken to, and the sting enters their souls.
>
> —*Women's Wages* (1893)

Clementina Maria Black (1853–1922)

Versatile is probably the best word to describe this Brighton-born woman: suffragist, writer, linguist, trade unionist, social reformer and family woman.

Clementina Black's circle included Marxist and Fabian socialists, such as Olive Shreiner and Eleanor Marx. She focused on the problems faced by working-class women and the nascent trade union movement. In 1886, Black was appointed honorary secretary of the Women's Trade Union League and moved an equal-pay motion at the 1888 Trades Union Congress. In 1889, she assisted in forming the Women's Trade Union Association, later the Women's Industrial Council.

Black could be found among the organisers of the Bryant and May match girls' strike in 1888. In 1895, she took on the editorship of *Women's Industrial News*, the journal of the Women's Industrial Council, which rallied middle-class women to research and report on the working conditions encountered by poorer women, and by 1914 had investigated around 120 trades. In 1896, she was campaigning for a legal minimum wage under the Consumers' League.

Unsurprisingly, by the early 1900s, Black was also moving in the women's suffrage campaign as secretary of the Women's Franchise Declaration Committee, which collected a petition of 257,000 signatures. Black joined the NUWSS as well the London Society of Women's Suffrage. By 1912–1913, she was editing *The Common Cause*, the 'organ of the women's movement for reform'.

Black's two political works, *Sweated Industry and the Minimum Wage* (1907) and *Makers of Our Clothes: A Case for Trade Boards* (1909) have been called 'powerful works of propaganda' (Grenier, 2004). Grenier tells us how 'Black's solution to the problem of low pay was the establishment of wages boards that would enforce a minimum wage for certain types of unskilled workers.'

Selina Cooper (1864–1946)

In 1876, aged 12, Selina Jane Cooper started work in the local textile mills at Barnoldswick near Skipton. She left school the following year and started work full-time in the mills. Cooper became active in trade union activities and took practical courses in laundry, hygiene and first aid. She joined the Women's Co-operative Guild in 1897 and the North of England Society for Women's Suffrage in 1900. That year, Selina Cooper and the North of England Women's Suffrage Society organised a petition signed by 29,000 women working in the Lancashire cotton mills and handed to Parliament on 18 March 1901:

(a) That in the opinion of your petitioners the continued denial of the franchise to women is unjust and inexpedient.
(b) In the home, their position is lowered by such an exclusion from the responsibilities of national life.
(c) In the factory, their unrepresented condition places the regulation of their work in the hands of men who are often their rivals as well as their fellow workers.

In 1901, Cooper was elected to the Board of Guardians, as a joint SDF-ILP candidate. The local paper, *The Colne and Nelson Times*, urged its readers not to vote for Selina Cooper; the arrant sexism is breathtaking:

> At the risk of being described ungallant, we have to ask the electors of Nelson to see to it that the three men candidates are elected Guardians ... We hold that the interests of women are not neglected by administrative bodies consisting entirely of men ... The three male candidates the ratepayers can elect ... have the leisure – without the domestic duties, which embarrass women.

Cooper lost patience with the SDF's complacency with the suffrage movement, and distanced herself from the party to become a full-time organiser for the suffrage movement. In 1910, she was one of four women selected to present the case for women's suffrage to Asquith, the then prime minister. During the First World War, Cooper opened the nation's first maternity centre in Nelson, Lancashire.

In November 1913, Selina Cooper and Margaret Aldersley addressed an open-air meeting in Haworth on women's suffrage. Selina's daughter, Mary, saw what happened next, not least her mother's amazing defiance:

> The men threw rotten eggs and tomatoes and all sorts of things ... we sheltered in a café. Mrs Aldersley went out and came back crying – covered with eggs and tomatoes ... My mother went out, and she said, 'I'm stopping here, whatever you throw, so go and fetch all the stuff you've got to throw, because,' she says, 'this blooming village would never have been known about but for three women – the Brontës.'

Women and birth control was another issue close to Cooper's heart, as testified by the following.

She assembled a library of books which she gave other women access to in the textile factory where they worked in the 1890s. One book was Dr Allison's *Book for Married Women* which offered this particularly emboldened statement:

> Women have rights as well as men, and to force a woman to have more children than her constitution will bear, or is her desire to have, is an act of cruelty that no upright man would sanction. It is against the true dignity of a woman to

> become a mere childbearing drudge. From a health standpoint, it is better to use preventative means.

Cooper's daughter recalls how her mother advised friends and neighbours on birth control:

> Birth control wasn't much talked about then, but my mother used to give a talk to the women that came here to our house about it ... She used to meet with a lot of opposition, because there were quite a lot of Catholics. But she used to talk about things – and elementary hygiene and elementary science and biology, and things we never got before.

Lady Constance Lytton

As the daughter of the Viceroy of India, Lady Constance Lytton (1869–1923) was born into a life of prodigious privilege. Her affinity with the suffrage movement saw her swap the luxury of palaces for the horror and indignity of prison cells. It was her campaigns for prison reform that led her join the suffragettes. In 1909, Lytton was locked up in Holloway but was swiftly released when prison officers discovered who she was. So Lytton took the alias Jane Warton, disguising herself as an 'ugly London seamstress' when doing activities that might land her in prison, such as pelting MPs' cars with stones. Lytton tried to carve the words 'Votes for Women' into her skin while in prison but only got as far as inscribing a letter 'V' over her heart when she was restrained by prison doctors. In 1914, she published her experiences of incarceration and force feeding, reflecting: 'When the ghastly process was over, I tapped on the wall and called out... "No surrender" and there came the answer past any doubt... "No surrender".'

Women of Colour Trailblazing in a Very White Male World

In the overwhelming majority of respects, women of colour who have won for themselves a reputation for trailblazing are no different at all from all women who have taken the initiative, been inspired and empowered, and who have succeeded as entrepreneurs and as leaders of men in all manner of industry and professions. So why a separate chapter? A trailblazer is a trailblazer, after all. The answer is because all of the women trailblazers in this book have had one almost insurmountable hurdle to negotiate that is not required of men and which men quite naturally are able to swerve around and preserve in their own interests. That hurdle of course is an obstacle placed in women's way by virtue of their sex. If this book has shown anything then it is the innate and enduring discrimination and sexism which prevailed before, during and after the Industrial Revolution, even to this day in some respects. Before a woman could succeed in very much anything apart from serial childbirth, childrearing, housework, selling umbrellas, and nimble-fingered factory work, she had to overcome the clannish, clubbish laddism which often insecure men clung onto in order to preserve their bastions of power and influence. The fact that so many women did just that is remarkable in itself, as clearly and impressively demonstrated by the 200 or so women celebrated in this book alone. There were of course many, many more women trailblazers out there who didn't make the book for reasons of page count and publishing economics.

Add to this the fact that you are also not white and your achievement is all the greater because being a woman in an overwhelmingly male dominion was hard enough; your skin colour made it harder still. While tribalism had been alive and well for centuries from the seventeenth and eighteenth centuries, racism as an independent social construct was beginning to take on more of the loathsome connotations it has today as a result of scrambling colonialism, arrogant entitlement, indented servitude and slavery, and assumed superiority over indigenous peoples. And that is why women of colour deserve a chapter all of their own.

Dido Elizabeth Belle (1761–1804)

Dido Elizabeth Belle was a British gentlewoman born into slavery in Pensacola in Spanish Florida, as the illegitimate daughter of a Royal Navy officer who later became an admiral – Sir John Lindsay. At the time he was captain of the British warship HMS *Trent*. Her mother was Maria Belle, an enslaved Black woman. Maria Belle was a 14-year-old child slave when she was taken, about the same time she fell pregnant by Lindsay, and gave birth to Dido when she was about 15 years old.[1] According to Paula Byrne, 'it is unlikely that the conception was consensual'.[2]

Lindsay took Dido (a popular name for slaves at the time, after the Queen of Carthage) with him when he returned to England in 1765, entrusting her upbringing to his uncle William Murray, 1st Earl of Mansfield, and his wife Elizabeth Murray, Countess of Mansfield. The Murrays educated Belle, raising her as a free gentlewoman at Kenwood House, together with another great-niece, the very British aristocrat Lady Elizabeth Murray (later Lady Elizabeth Mary Finch-Hatton), whose mother had died, making Lady Elizabeth and Belle second cousins. Belle lived comfortably at Kenwood for thirty-one years. In his will of 1793, Lord Mansfield bequeathed to her £500 and a £100 annuity. In 1799, Belle also inherited £100 from Lady Margery Murray, one of two unmarried aunts who had previously come to live with the Murrays as carers in their later years.[3]

Belle also received an annual allowance of £20, plus an additional £5 on her birthday and at Christmas. By contrast, Lady Elizabeth received £100, excluding her birthday and other gifts. Dido's allowance was also paid quarterly – £5 every three months – while Lady Elizabeth received £50 every six months. For comparison, the annual wage of a female housekeeper in a top-drawer household ranged from £20 to £70 at that time, while a lieutenant in the Royal Navy would take about £100 a year. About £200 could buy you a three-bedroom house with a garden outside the city of London.

At Kenwood, it would seem that Belle became more of a lady's companion than a lady's maid. The Mansfields treated her well and brought her up to be an educated woman. As she grew older, she assisted Mansfield by taking dictation of his letters. In addition, Dido Elizabeth Belle would work as manager in the dairy and poultry yard, a typical occupation for ladies of the gentry; helping her uncle with his correspondence was less usual. English Heritage tells that 'the idea of a biracial child born in this era to be raised as part of an aristocratic British family was virtually unheard of'. Byrne adds 'aspects of Belle's life, such as being given expensive medical treatments and luxurious bedroom furnishings, were evidence of her position as Lady Elizabeth's equal at Kenwood'.

A brief reference to Belle occurs in volume II of James Beattie's *Elements of Moral Science*. Beattie refers to her intelligence, saying:

> But I happened, a few days after, to see his theory overturned, and my conjecture established by a negro girl about ten years old, who had been six years in England, and not only spoke with the articulation and accent of a native, but repeated some pieces of poetry, with a degree of elegance, which would have been admired in any English child of her years.

Notes

1. Her age was confirmed by the Pensacola property record about her later life: 'the manumission transaction for the sum of two hundred Spanish milled dollars paid by Maria Belle a Negro Woman Slave about 28 years of age', dated 22 August 1774; this confirmed that Maria Belle was about 14 when Dido was conceived.
2. Paula Byrne, *Belle: The True Story of Dido Belle*, Harper Audiobooks, 2014.
3. Historic England, 'Slavery and Justice at Kenwood House, Part 1' (PDF) tell us from a contemporary obituary that '[Lindsay] has died, we believe, without any legitimate issue but has left one natural daughter, a Mulatta who has been brought up in Lord Mansfield's family almost from her infancy and whose amiable disposition and accomplishments have gained her the highest respect from all his Lordship's relations and visitants.'

Sara Forbes Bonetta (Aina) (1843–1880)

Much of the following is derived from Akira Chiba's *Queen Victoria and the African Princess*.

Sara or Sally Forbes Bonetta (born Aina or Ina) was ward and goddaughter of Queen Victoria. She was a titled member of the Egbado clan of the

Yoruba people in West Africa, who was orphaned during a war with the nearby Kingdom of Dahomey as a child, and was later taken into slavery. Sara was then gifted to Captain Frederick E. Forbes of the Royal Navy, who was in West Africa on a British diplomatic mission, where he unsuccessfully attempted to negotiate with King Ghezo to end Dahomey's participation in the Atlantic slave trade. As was the custom, Forbes and King Ghezo exchanged gifts with each other. King Ghezo offered Forbes a footstool, rich country cloth, a keg of rum, ten heads of cowry shells, and a caboceers stool. Aina became a goddaughter of Queen Victoria. She was later commanded by the Queen to marry Captain James Pinson Labulo Davies, a wealthy Lagos philanthropist.

How on earth did all this come to be?

King Ghezo was adamant he would not stop his trade in slaves. According to Bethany Kujawinski, Forbes heard a scream and then saw a group of Dahomans who were waving their guns and carrying people in little baskets. Forbes was informed by an interpreter that those in the baskets were going to be executed. The ceremony was called the 'Ek-onee-noo-ah-toh' or 'watering of the graves'.

Walter Myers takes up the shocking story:

> The people in the baskets were dressed in white garments, were to be slaughtered and their blood dripped on the graves of high ranking Dahomans. Some of the intended victims had been held in captivity for over two years for this tradition. They were being carried while their hands and feet were tied together. As the victims were dragged through the ranks, the Dahomans poked and jabbed them with knives and spears.

Commander Forbes watched in horror as a man from the basket tipped over to a pit and the man viciously fell down. As he hit the ground, he was instantly attacked and his head cut off. Forbes tried but failed to make King Ghezo stop the ritual. Forbes then offered him money. The King eventually allowed him to bargain for some of the victims. However, King Gezo's interpreters clarified that the custom of watering the ancestors' graves was an ancient one and could not be discontinued without dishonouring the Dahomey people.

Myers continues in his disturbing report, saying that Forbes had never witnessed a ritual this vicious and violent. He was certain that this was the worst moment he had ever encountered in all the years fighting the slave trade. He then noticed the girl, Aina. She was so tiny, so still. As they carried her closer to the pit, the drums became more intense. Forbes was appalled. He found it hard to comprehend how a king could ritually murder a child. However, Gezo found it extremely easy to sacrifice the girl. It was explained that she was an Egbado, a Dahoman enemy. Her blood on the king's ancestors' tombs would be a tribute to them.

Forbes panicked and assured King Ghezo that Queen Victoria would not honour a king that would kill a child, so the king offered Aina to be a gift for Queen Victoria. Forbes estimated that Aina had been enslaved by King Ghezo for two years. Although her actual ancestry is unknown, Forbes came to the conclusion that Aina was likely to have come from a high-status background because of the tribal markings on her face and the fact that she had not already been sold to European slave traders.[1] Describing Aina in his journal, he wrote: 'one of the captives of this dreadful slave-hunt was this interesting girl. It is usual to reserve the best born for the high behests of royalty and the immolation on the tombs of the deceased nobility...'[2]

English Heritage informs us that Dahomey was notorious for mass executing its captives in spectacular human sacrifice rituals as part of the Annual Customs of Dahomey. Forbes was aware of Aina's potential deadly fate in Dahomey, and as he wrote in his journal, refusing Aina 'would have been to have signed her death-warrant, which probably would have been carried into execution forthwith'. Captain Forbes accepted Aina on behalf of Queen Victoria and embarked on his journey back to Britain.

Forbes renamed her Sara Forbes Bonetta, after himself and his ship HMS *Bonetta*. He initially intended to raise her himself. However, Sara was later taken to meet Queen Victoria and she told her majesty about her horrible life, that for a couple of years she was kept in a small cage next to other unfortunate prisoners who she watched from time to time be taken out and slaughtered in rituals of the King of Dahomey.

After the wedding to James Pinson Labulo Davies, the couple moved back to Africa, where they had three children: following the births, she began work as a teacher. Sara continued to enjoy such a close relationship with Queen Victoria that she and Bishop Samuel Ajayi Crowther were the only Lagos *indigènes* the Royal Navy had standing orders to evacuate in the event of an uprising in Lagos.

Mary Prince

Some Black slave immigrants came with their owners from North America and the Caribbean, brought to work as servants in various households. A good example of this was Bermudan slave Mary Prince (1788–after 1833), who was forcibly shipped here in 1828 but was able to escape her owner with the help of white servants. This is recounted in her autobiography, the first book published in Britain by a Black female author.

Like most slaves, Mary Prince was illiterate, but while living in London she showed great initiative when arranging to dictate her life story to Susanna Strickland, a lady living in the home of Thomas Pringle, secretary of the Society for the Mitigation and Gradual Abolition of Slavery Throughout the British Dominions (aka Anti-Slavery Society, 1823–1838). Strickland transcribed Prince's life story which was published as *The History of Mary Prince* in 1831, the first account of the life of a Black enslaved woman to be published in the United Kingdom (Mary Seacole's wonderful *Wonderful Adventures of Mrs. Seacole in Many Lands* was published in 1857). Prince's first-hand description of the brutalities of enslavement, published at a time when slavery was still legal in Bermuda and British Caribbean colonies, did much to galvanise the British anti-slavery movement. It was reprinted twice in its first year.

Mary Seacole is covered in the chapter on nursing above; Cornelia Sorabji is covered in the Medical Doctors chapter.

Notes

1. *Picture World: Image, Aesthetics, and Victorian New Media*, Rachel Teutolsky, Oxford University Press, 2020, p. 267.
2. Akira Chiba, *Queen Victoria and the African Princess*, Museion Books.

See also Dorothy Thomas (1756–1846) and Mary Seacole (1805–1881) above.

Landowners

Anne Lister (Gentleman Jack), the doyenne of female landowners of the time, is covered in my *Bridgerton Revisited* (2026).

Fanny Talbot (1824–1917)

Fanny Talbot was a landowner and philanthropist, and a friend of social reformer, philanthropist and art critic John Ruskin (1819–1900). She is famous for donating the first property – 4.5 acres of land known as the Cliff of Light or *Dinas Oleu* at Barmouth, Gwynedd – to the National Trust, as founded by Hardwicke Rawnsley, Robert Hunter and Octavia Hill and others.

In 1874, Talbot made Ruskin an offer for the Guild of St George, to donate eight cottages and a 4.5-acre area of land at Barmouth.[1] Margaret Spence tells how her generous offer surprised Ruskin and their friendship flourished thereafter. The Guild of St George is a charitable English Education Trust, but with a global membership, which tries to uphold the values and implement the ideas of John Ruskin.[2]

This is how Ruskin described Fanny Talbot:

> She's a motherly, bright, black-eyed woman of fifty with a nice married son who is a superb chess player. She herself is a very good one, and it's her greatest indulgence to have a written game with me. She's an excellent nurse, and curious beyond any magpie that ever was, but always giving her spoons away instead of stealing them. Practically clever beyond most women; but if you answer one question she'll ask you six![3]

In 2001, the Guild's collection moved to the Ruskin Gallery in Sheffield's new Millennium Galleries. In 2011, the gallery was renamed the Ruskin Collection.

Notes

1. Blanche Atkinson, *Ruskin's Social Experiment at Barmouth*, Sothis Press, 2023.
2. Margaret E. Spence, 'Ruskin's Friendship with Mrs Fanny Talbot', *Bulletin of the John Rylands Library Manchester*, 1960.
3. Again, according to Margaret Spence.

Lawyers

Eliza Orme (1848–1937)

Eliza Orme was the first woman to earn a LLB degree (Hons) in England, from University College London in 1888. The seventh of eight children, her father Charles Orme was a distiller and Eliza Orme, her mother, hosted a salon that was frequented by Pre-Raphaelites. Both parents were supportive of women's education. Orme's sister Emily Rosaline Orme became a notable campaigner for women's suffrage in Scotland.

So, the environment was right for Eliza to listen and learn: academics and artists such as Thomas Carlyle and John Stuart Mill visited; Orme studied at the Bedford College for Women and, in 1869, was one of nine women to write the University of London's first General Examination for Women. When University College London started to allow women as well as men to attend lectures, she enrolled as a student there in 1872. She won awards and scholarships in Political Economy, Jurisprudence, and Roman Law.[1]

In 1874, she wrote two articles for *The Examiner* on degrees for women at the University of London, arguing strongly in favour of women's higher education. Orme began working towards legal practice in 1872, helped by Helen Taylor, who paid her fee to enable her to become a pupil at Lincoln's Inn in the chambers of barrister John Savill Vaizey in 1873. However, her ambition to be recognised as a 'conveyancer under the bar' was blocked – until the Sex Disqualification (Removal) Act 1919, women could not qualify either as barrister or solicitor in England. Instead, Eliza set up an office on Chancery Lane in 1875.

Her work here involved wills, mortgages and property conveyancing; patents and probate settlements could be done by non-solicitors as they were unregulated.

In 1893, Orme was invited to send papers to the Congress on Jurisprudence and Law Reform as part of the World's Columbian Exposition in Chicago

– the first time women participated in a formal legal congress. Her paper, *The Legal Status of Women in England*, was read by Mary A. Ahrens.

She was active in Liberal Party politics, the National Society for Women's Suffrage and the Society for the Promotion of the Employment of Women, and other feminist organisations. Orme was a founding member of the Women's Liberal Federation in 1887 and editor of the *Women's Gazette and Weekly News* between 1889 and 1891.

Notes

1. Mary Jane Mossman, 'Introduction: The First Women Lawyers', *The First Women Lawyers: A Comparative Study of Gender, Law and the Legal Professions*, Hart Publishing, 2006.

Letitia Alice Walkington (1857–1918)

Letitia Alice Walkington was born in Belfast, the daughter of Letitia (née Von Heyn) and Thomas R. Walkington. Her father was a descendant of Anglican Bishop Edward Walkington and her mother was the daughter of Prussian consul and Belfast ship owner, Gustav von Heyn; in 1913 Gustav Wilhelm Wolff of Harland & Wolff left both Edith, her sister, and Letitia £200. Letitia did not go to school until she was 16, home educated as she was by a governess. Then she went to boarding schools in England and Paris.

She was the first woman to graduate with a degree of Bachelor of Laws in Great Britain or Ireland, which she received from the Royal University of Ireland (RUI) in Dublin. In 1882 Walkington had passed the matriculation exam to go there; this was prior to the Irish Universities Act 1908 so women who had passed the matriculation exam for the Royal University could also take classes at Queen's College, Belfast. She completed her BA in 1885 and her MA in 1886, both in the area of Logic, Metaphysic, and Political Economy. She then completed her LL.B. in 1888 and her LL.D in 1889, becoming the first woman to complete these degrees.

Despite several offers of chamber work, she opted instead to coach other young women for examinations as there were limited opportunities for them in the law schools. Walkington also worked on a device for embossing Braille.

She was active in both the suffrage and the temperance movements. In 1889, Walkington participated in the 'Congrès International des Oeuvres et Institutions Feminines' in connection with the Paris Exhibition as part of the Dublin Women's Suffrage Committee. She co-founded the Belfast Women's Suffrage Society in 1912, which became the Women's Political League in 1918, and helped set up the Irish Women's Suffrage Federation. She also served as secretary for the Belfast chapter of the Church League for Women's Suffrage and worked with the Irish Women's Temperance Union.

Cornelia Sorabji (1866–1954)

Cornelia Sorabji was the first female graduate from Bombay University (with a first class in Literature) and the first woman to take the Bachelor of Civil Laws (BCL) exam at Oxford in 1892. Sorabji spent her adult life fighting to improve education for and legal representation of women and children in India. She was a student at Somerville Hall (later College) and was tutored by the eminent classicist, Benjamin Jowett – as vice chancellor he enabled her to sit the BCL exams with the other students in the Examination Schools.

Sorabji became the first woman to practise law in India as the first female advocate, operating for many years as a legal advisor to women and children on cases in provincial courts, and as a 'Lady Assistant' to the Court of Wards. She finally became qualified as a barrister when women were admitted to the profession for the first time in 1923.

She was involved with a number of pressure groups, including:

- The National Council for Women in India
- The Federation of University Women
- The Bengal League of Social Service for Women

Her mother, Francina Ford-Sorabji (née Santya), was adopted aged 12 and raised by a British couple; she helped set up girls' schools in Poona. Her mother's support for girls' education, and care for the local disadvantaged, was an inspiration for Cornelia Sorabji to advocate for women. Francina Ford encouraged her students, and her seven daughters, into higher education and the professions, including law, medicine, and midwifery: Cornelia apart, there was also educationalist Susie Sorabji, and medical doctor Alice Maude Sorabji Pennell. Francina Sorabji also ran a teacher-training programme; she went to England to fundraise for her work in Poona in 1886, and testified before a British commission on education in India. She fostered orphans and welcomed widows and their children into her home. During an outbreak of plague in 1896, she helped to introduce preventive public health in villages around Poona.

Sorabji had an ambivalent and cautious attitude to British rule: she was careful about transplanting Western values to India and did not support Gandhi's campaign for Indian self-rule. In short, she was an advocate for women and children and social reform, banging the drum especially for education for girls, for legal rights for women subjected to segregation (*purdahnashins*) and forbidden to communicate with the outside male world, for abolition of child marriage, and for protection and support for widows. Often, *purdahnashins* women owned considerable property, yet they had no access to legal expertise to protect it. Sorabji was given special dispensation to enter pleas on their behalf.

Sorabji was a prolific author; her earlier books include:

- 1901: *Love and Life beyond the Purdah* (London: Fremantle & Co.)
- 1904: *Sun-Babies: Studies in the Child-life of India* (London: Blackie & Son)
- 1908: *Between the Twilights: Being Studies of India Women by One of Themselves* (London: Harper)
- 1916: *Indian Tales of the Great Ones Among Men, Women and Bird-People* (Bombay: Blackie)
- 1917: *The Purdahnashin* (Bombay: Blackie & Son)

She later edited *Queen Mary's Book for India* (Harrap & Co., 1943), which had contributions from such authors as T. S. Eliot and Dorothy L. Sayers. She contributed to a number of periodicals, including *The Asiatic Review*, *The Times Literary Supplement*, *Atlantic Monthly*, *Calcutta Review*, *The Englishman* Magazine, *The Statesman* and *The Times*.

Educators

Harriet Martineau (1802–1876)

Norwich-born Harriet Martineau was a prolific writer on topics from philosophy to travel, from slavery to American society. She wrote from a sociological and feminist point of view and was one of our most influential abolitionists.

Her life is characterised by a lack of affection between her and her mother, instigated by the hiring of a wet nurse whose lactation was insufficient, causing Harriet much distress. Her troubled childhood is examined in her *Household Education* (1848). Her mother, Mrs Martineau, strictly enforced what was regarded as proper feminine behaviour, pushing her daughter to 'hold a sewing needle' as well as the (hidden) pen.[1]

Home schooling was the order of the day in the Martineau household: Harriet was taught by several of her elder siblings and was taught French by her mother, which was her father's first language. He taught her Latin and her brother Thomas taught Harriet maths and writing. Being taught at home, especially by all her siblings, often led to mockery.[2] Aged 9, Harriet graduated to a small school run by a man named Perry which she says was 'the catalyst for her intellectual development and interest in education'.

According to Hoecker-Drysdale:

> As her education progressed she grew fond of the subjects of Shakespeare, political economy, philosophy and history; despite her love for these topics, her mind was often dominated by the three biggest insecurities in her life: her hearing disability, her poor handwriting, and the look of her hair.

Harriet Martineau then moved to a girls' boarding school run by her aunt and uncle in Bristol where she independently studied Latin, Greek, Italian, and focused more on the Bible.

As noted, chronic illness dogged Martineau: she started losing her senses of taste and smell when quite young. She was deaf and used an ear trumpet at the age of 12 but neglected it most of the time it until her late 20s in order to avoid the bullying which came her way as an adolescent with an obvious disability. With her illnesses and the later death of her father, she had no choice but to make a living for herself so she began to write anonymously for the *Monthly Repository*, a Unitarian periodical. Her first contribution was 'Female Writers of Practical Divinity', and in 1823 she published *Devotional Exercises and Addresses, Prayers and Hymns*.

Her earliest novels were published beginning with *Principle and Practice* in 1827 and *Five Years of Youth: or, Sense and Sentiment*, in 1829.

Her father died in 1826, and in 1829, when the family's textile business failed, Martineau, then 27 years old, escaped the traditional roles of feminine propriety and social constriction to earn a living for her family: as well as her needlework her regular work with the *Repository* helped establish her as a dependable and popular freelance writer.

Martineau says in her *Autobiography* that her father's business failure was 'one of the best things that ever happened to us'. She described how she could then 'truly live instead of vegetate' while she wrote '[her] fusion of literary and economic narratives'.

Harriet's first commissioned book, *Illustrations of Political Economy*, was a fictional tutorial designed to help the general readers grasp the ideas of Adam Smith. *Illustrations* was published in February 1832 in a short edition but it soon became a bestseller and would out-sell the books of Charles Dickens. *Illustrations* was her first work to receive wide acclaim, and its success served to disseminate the free-market ideas of Adam Smith and others throughout the British Empire.

Later works provided fictional tutorials on political economists such as James Mill, Jeremy Bentham and David Ricardo. Martineau was a follower of Malthus and in stories such as 'Weal and Woe in Garvelock', she promulgated population control through Malthus's voluntary checks' – voluntary chastity and delayed marriages.

In 1834–36, Martineau embarked on a lengthy trip to the United States, meeting many abolitionists and studying the emerging schools for girls' education. Her support of abolitionism, then widely unpopular across the

US, caused controversy and was stoked up by her *Society in America* (1837) and *How to Observe Morals and Manners* (1838). In the former, Martineau criticised the poor state of women's education. She wrote:

> The intellect of women is confined by an unjustifiable restriction of … education … As women have none of the objects in life for which an enlarged education is considered requisite, the education is not given … The choice is to either be ill-educated, passive, and subservient, or well-educated, vigorous, and free only upon sufferance.[3]

Martineau's *Illustrations of Political Economy* continued to sell well; 'by 1834, the monthly sales … had reached 10,000 in a decade in which a sale of 2,000 or 3,000 copies of a work of fiction was considered highly successful.'[4]

Martineau's meeting with Charles Darwin in 1836 produced a strange letter to Erasmus Darwin (his brother) describing the encounter in which inappropriate comments are made and Martineau is subjected to the defeminisation by men so common at the time:

> She was very agreeable and managed to talk on a most wonderful number of subjects, considering the limited time. I was astonished to find how little ugly she is, but as it appears to me, she is overwhelmed with her own projects, her own thoughts and own abilities. Erasmus palliated all this, by maintaining one ought not to look at her as a woman.[5]

From 1852 to 1866, she contributed regularly to *The Daily News*, sometimes writing six leaders a week. She wrote more than 1,600 articles for the paper in all. The paper was originally founded and edited by Charles Dickens from 1846 as rival to the right-wing *Morning Chronicle*.

She taught that study of the society must embrace all aspects, including major political, religious and social institutions, and she insisted on the imperative need to include the lives of women. Unbelievably, she was the first sociologist to study such issues as marriage, children, religious life, and race relations, urging her fellow sociologists to act and not just observe but to work to the benefit of society. Surely another sign that supposedly 'women's' subjects were given short shrift in academia and by politicians.

Notes

1. Diana Postlethwaite, 'Mothering and Mesmerism in the Life of Harriet Martineau', *Signs*, 14(3), pp. 583–609, 1989.
2. Susan Hoecker-Drysdale, *Harriet Martineau: The First Woman Sociologist*, Berg, 1992.
3. Postlethwaite, 1989.
4. Elaine Freedgood, 'Banishing panic: Harriet Martineau and the popularization of political economy', *Victorian Studies*, 39(1), pp. 33–53, 1995.
5. 'Letter 325; Darwin, C. R. to Darwin, C. S. (7 Dec 1836)'. Darwin Correspondence Project.

Margaret Byers (1832–1912)

> '[A] strong supporter of women's academic abilities and a pioneer of their academic rights.'

Margaret Byers (née Morrow) was an eminent Irish educator, activist, social reformer, missionary, and writer. She founded Victoria College in Belfast, a pioneer establishment in education, and was deeply involved in philanthropic work, focusing on the training of the young. She wrote many papers on girls' education in Ireland, on Irish industrial schools, and on temperance. Relocating to Belfast in 1859, she established the Ladies' Collegiate with pupil numbers rising from thirty-five in 1859 to more than sixty within a few years. In 1887, the school was rebadged as Victoria College and School, where Byers broke with tradition, offering her female students academic courses, for example in history, philosophy and science, as well as Latin and Greek (still necessary for university entrance). Significantly, the students were trained to take up positions as governesses or teachers with highly qualified staff ever raising standards and building respect for female academic achievement.

Along with the educational and political activist Isabella Tod of the North of Ireland Suffrage Society, and others, in 1874 Byers founded the Belfast Women's Temperance Association and Christian Workers' Union. From this came the Belfast Prison Gate Mission for Women and the Victoria Homes for the Reclamation and Training of Neglected and Destitute Girls.

Byers worked with Tod to further advance educational opportunities for young women. Both were active in founding the Ladies Institute, which advocated female education, in 1867, as well as the Ulster Schoolmistresses Association. She backed Tod in her successful campaign to have girls included in the 1878 Education (Ireland) Act, enabling them to sit public exams, and in the 1879 Universities (Ireland) Act which entitled women to be granted degrees from the Royal University of Ireland.

She was awarded an LLD from Trinity College Dublin in 1905, the first Ulsterwoman to be awarded an honorary degree from any university, and in 1908, she was appointed to the Senate of the new Queens University of Belfast.

Margaret Byers also helped establish industrial schools for poor girls in Belfast and the north and an active suffragist and committee member of the Belfast Liberal Unionist Association.

Elizabeth Corbet Yeats (1868–1940), aka Lolly

Lolly was an eminent Irish educator and publisher. She was an art teacher and author of several books on art, and was a founder of Dun Emer Press, publisher of several works by her brother W. B. Yeats.

In 1908, Lolly and her brother William set up the Cuala Press, publishing over seventy books, including forty-eight by the poet. The Press only employed women, supported women's rights, and 'promoted a new national identity and confidence after the 1921 independence of Ireland from Britain'.

Mary Hare (1866–1945)

The remarkable Mary Hare established a uniformed women's police force in Brighton and was a restless suffragette. However, her crowning achievement was the founding of an oral school for deaf children which is today the UK's largest non-maintained special school for the deaf.

Mary Hare's unique ethos was that 'deafness was not a mental disability but rather a sensory impairment that presents the deaf child with additional barriers to learning'. To this day, in the twenty-first century, the school prolongs her vision for the auditory/oral education of deaf children. Mary Hare wrote in her will, 'my efforts on behalf of the Deaf have been my greatest joy in life'.

Mary opened her school for deaf boys and girls in her mother's house in London in 1883. She went on to establish Dene Hollow Oral School for the Deaf in 1916, taking pupils from all over the world.

Philanthropists

Faith Gray (1751–1826)

A diarist, philanthropist and school founder who, with Catharine Cappe, improved education in York for poor girls at the Grey (now Blue) Coat School.

Faith was the first of seven children; in 1765, aged 14, she began the diaries for which she is remembered, recording how she spent her time reading and sewing with her mother while her brothers read and learnt languages.

The year 1782 saw her begin an all-important project with Catharine Cappe: the pair organised evening classes to teach workers at a local hemp factory to read, and at Sunday school each week. Two years later, with Cappe, Gray established a School for Spinning Worsted in York, offering an education for girls. The school was staffed by women volunteers. Cappe wrote that the purpose was to 'excite a spirit of virtuous industry among the children of the poor'. The parents were all for it because the girls were clothed – they were taught how to make garments from the yarn they had spun.

Given this education, the students benefited by being able to apply for better-paid work and avoid the unsatisfactory options usually open to poor poorly uneducated girls.

Gray was then approached to become involved with the Grey Coat School in York which was in the doldrums. She wrote notes to the (male) governors of the school to help them out. The master of the school was deemed unfit and his wife had poor mental health. The Ladies' Committee was permitted to take over management of the school.

Faith Gray founded the York Female Friendly Society in 1788, open to former students of the Grey Coat School and to their 'School for Spinning Worsted'.

Angela Burdett-Coutts (1814–1906)

Angela Georgina Burdett-Coutts, 1st Baroness Burdett-Coutts (née Burdett) was the doyenne of female philanthropy; her list of good causes is as impressive as it is diverse. She was the daughter of Sir Francis Burdett, 5th Baronet, and Sophia, formerly Coutts, daughter of banker Thomas Coutts. Eighteen thirty-seven was a very good year for Burdett-Coutts: she became one of the richest women in England when she inherited her grandfather's fortune of around £1.8 million (£210,000,000 in 2023). Edward VII described her as 'after my mother, the most remarkable woman in the kingdom'.

Burdett-Coutts spent part of each year at Ehrenberg Hall in Torquay with her former governess and later devoted companion Hannah Brown. On Brown's passing in 1878, Burdett-Coutts confided to a friend that she was utterly crushed by the loss of 'my poor darling, the companion and sunshine of my life for 52 years'. She was a friend of both Charles Dickens and the Duke of Wellington, to whom she proposed marriage despite the great difference in their ages.

Sixty per cent of her income went to her sister in 1881 when she was 67 and shocked polite society by marrying her 29-year-old secretary, the American-born William Lehman Ashmead Bartlett, who became MP for Westminster that year; but because her husband was American by birth, a clause in her step-grandmother's will forbidding her heir to marry a foreign national was invoked.

Nevertheless, Burdett-Coutts spent most of her wealth on scholarships, endowments, and a huge range of philanthropic causes. Here are some of them, just in the UK:

- In 1847 she co-founded with Charles Dickens a home for young women who had 'turned to a life of immorality', mainly petty theft and prostitution. This was Urania Cottage, a reformatory, which has been called a 'discreet version' of London's Magdalen House for Reception of Penitent Prostitutes.[1] In an anonymous article published in *Household Words* in 1853, Dickens called it a 'Home for Homeless Women'.[2]
- She was particularly interested in improving the lot of indigenous Africans, and in the education and relief of the poor or suffering generally.
- Burdett-Coutts was a significant benefactor of the Church of England, building two churches – one of which was St Stephen's Church, Rochester Row, Westminster – and endowing church schools.

- She built a primary school in Westminster known as Burdett-Coutts & Townshend Foundation Church of England Primary School.
- She got involved with the Royal Society for the Prevention of Cruelty to Animals (RSPCA), helped in the formation of the RSPCA's ladies committee in 1869 and was its first president.
- Burdett-Coutts founded Columbia Market in 1869 in Bethnal Green. Her project in Columbia Square led to her becoming a pioneer in social housing. Her small housing development, Holly Village, on the corner of what was then her estate, is now in private hands in Highgate. It was the first gated community, in 1865, and is a Grade II-listed site.
- She was the president of the British Goat Society and President of the British Beekeepers Association 1878–1906.
- She founded the Westminster Technical Institute in 1893.
- She funded church bells for St Paul's Cathedral.
- She supported the London Ragged School Union.
- She established a sewing school for women in Spitalfields, when the bottom fell out of the local silk trade.
- She promoted the fishing industry in Ireland by helping to start schools and provide boats; she also advanced £250,000 in 1880 for supplying seed potatoes to the impoverished tenants.
- She arranged for the placement of hundreds of destitute boys in training ships for the navy and merchant service.
- She supported the British Horological Institute.
- She supported the Cotton Supply Association and contributed £500 in 1870 to the organisation.
- She purchased more than 100 New Testament Greek manuscripts from a dealer in Janina in Epirus in 1864, shipped them to England between 1870 and 1872, and presented them to Sir Roger Cholmeley's School, Highgate.
- She funded the Burdett-Coutts Memorial Sundial, commissioned for St Pancras Old Church, listing the names of many people whose bodies had been dug up from the churchyard to make space for the Midland Railway.

Notes

1. Claire Wood and Juliet John, *Edinburgh Companion to Charles Dickens and the Arts*, Edinburgh University Press, 2024, p. 16, note 25.
2. Lillian Nayder, *The Other Dickens: A Life of Catherine Hogarth*, Cornell University Press, 2012, p. 148, note 25.

Enriqueta Rylands (1843–1908)

The John Rylands Research Institute and Library is one of, if not the, most beautiful buildings in Manchester: this masterpiece of Victorian Gothic architecture is more like an ornate castle or cathedral.

Architecture apart, highlights include 'the oldest known piece of the New Testament, the St John Fragment, magnificent illuminated medieval manuscripts and a 1476 William Caxton edition of Chaucer's *Canterbury Tales*'.

We owe it all to the devoted, prescient and perceptive wife of John Rylands, Enriqueta. She was determined to give permanence and an eternal legacy to her husband's huge contribution to Manchester and Manchester's cultural life and so she bought some land on what is now Deansgate and built this monumental depositary for the some of the most important literary works in the land.

Enriqueta based her collection on the Althorpe library which she purchased from the 2nd Earl Spencer and built it up magnificently as a start to the wonderful library we see today. Exquisitely furnished and appointed with shades of an Oxbridge college, the John Rylands Library is surely one of the jewels in the crown, not just of Manchester's splendid libraries but those of Britain and, yes, the world.

The 250,000 (and growing) volumes and one million-plus archival items include rare manuscripts by Caxton and the personal archives of Elizabeth Gaskell and John Dalton.

Animal Welfare

Emily Williamson (1855–1936)

Emily Williamson is another significant woman who has been silently airbrushed out of the history books until quite recently. Her long-overdue statue was unveiled in Didsbury's Fletcher Moss Park on the anniversary of her birth on 17 April 2023.

You have to take your hat off to Emily Williamson, not only was she a fervent suffragist but she was also opposed to the use of feathers as exhibited in the silly hat creations of the day. Emily was determined to put an end to this, mindful of the death toll in birds which was the collateral damage required to sport such pointless plumage, all in the name of one-up(wo) manship and vanity.

What was the trigger? Emily was given to watching one of her favourite species, the Great Crested Grebe, being hunted to extinction; she implored the British Ornithologists' Union to take a stand against 'murderous millinery', but they ignored her letters.

Emily set the ball rolling when she shared her disgust at this appalling trade with some like-minded friends at her home, the Croft, in Fletcher Moss Gardens. She launched a campaign to put an end to it all when she invited them to sign a pledge to wear no feathers. And so, the Plumage League was hatched, later to become the revered Society for the Protection of Birds in 1889 with 20,000 members in the first ten years, made up of over 150 branches and doubling by 1898. Queen Victoria became an early member, which may well have given pause for thought to the inevitable largely male scoffers.

In 1891, Emily joined forces with another all-female group – the Fur, Fin and Feather Folk of Croydon. Disaffected milliners, bird killers, fashionistas and posh women were outraged, dubbing the RSPB 'feather faddists'. Explicit exceptions included birds killed for food and the ostrich, because the harvesting of its tail feathers was not painful. Initially, all members were women. It took thirty years for the Plumage Importation (Prohibition) Act to be passed, bringing an end to the trade in exotic bird skins – and endangered species began to recover.

But Emily didn't stop there: in 1891, she founded the Gentlewomen's Employment Association in Manchester, which in turn spawned first the Princess Christian Training College for Nurses, pioneering the use of real living infants instead of dolls. Another first followed in 1898 with the Loan Training Fund, which subsidised further education costs for young women, the first such scheme in the UK.

Frances Power Cobbe (1822–1904)

An Anglo-Irish writer, philosopher, religious thinker, social reformer, traveller, and anti-vivisection activist, Cobbe was a woman with many strings to her bow: she founded a number of animal welfare groups, including the National Anti-Vivisection Society (NAVS) in 1875 and the British Union for the Abolition of Vivisection (BUAV) in 1898. She sat on the executive council of the London National Society for Women's Suffrage.

In 1857, the death of Cobbe's father and the associated annuity enabled her to travel on her own around parts of Europe and the Near East. In Italy, she joined with a commune of similarly independent women: Isa Blagden, with whom she went on briefly to share a house, the sculptor Harriet Hosmer, the poet Elizabeth Barrett Browning, the painter Rosa Bonheur, the scientist Mary Somerville and the Welsh sculptor who became her partner, Mary Lloyd, to whom she in letters and published writing variously referred to as 'husband', 'wife', and 'dear friend'.

When focused on writing, she began to publish in Victorian periodicals and was able to support herself by writing. She became involved in feminist campaigns for the vote, for women to be admitted to study at university on the same terms as men, and for married women's property rights.

Her 1878 essay *Wife-Torture in England* influenced the passing of the 1878 Matrimonial Causes Act, which gave women subjected to abusive husbands the right to a legal separation.

Cobbe applied her moral theory to animal rights, first in *The Rights of Man and the Claims of Brutes*, 1863. She argued that humans may do harm to animals in order to satisfy real wants but not from mere 'wantonness'. For example, humans may eat meat but not kill birds for feathers to decorate hats. The harm or pain inflicted must be the minimum possible. For Cobbe this set limits to vivisection, for example, it must always be done under anaesthesia.

Her many books included: *The Pursuits of Women* (1863), *Darwinism in Morals, and other Essays* (1872), *The Hopes of the Human Race* (1874), *The Duties of Women* (1881), *The Peak in Darien, with some other Inquiries touching concerns of the Soul and the Body* (1882), *The Scientific Spirit of the Age* (1888) and *The Modern Rack: Papers on Vivisection* (1889), as well as her autobiography.

Philosophers

Harriet Taylor Mill (1807–1858)

Harriet Taylor Mill was an English philosopher and women's rights activist. Harriet has been overshadowed in history by her eminent husband and philosopher John Stuart Mill but, as a philosopher in her own right and an influence on her spouse, she deserves to be recognised as the individual she was and is.

The *Stanford Encyclopedia of Philosophy* begins its entry on Harriet Taylor with a caveat:

> Harriet Taylor Mill (1807–1858) poses a unique set of problems for an encyclopedist. The usual approach to writing an entry on a historical figure, namely presenting a straightforward summary of their major works and then offering a few words of appraisal, is not available. This is because she worked in such close collaboration with John Stuart Mill that it is exceedingly difficult to disentangle their contributions to the products of their joint effort. In attempting to assess Taylor Mill's philosophical career, one encounters sharply conflicting reports about her ability from people who knew her, contradictory evidence about her role in authoring several books and essays, and widely varying judgments about how much influence she exerted on Mill's thought and work.

So, with that in mind… In 1826, she married John Taylor, a prosperous merchant; together they became active in the Unitarian Church and in 1830 a Unitarian minister introduced Harriet to the philosopher John Stuart Mill. An affair ensued which was to last for more than twenty years – in the early days, the intimacy was more or less tolerated by Harriet's husband: John Mill visited the Taylors' home most nights, visits that John Taylor would accommodate by going out to his club.

From 1833, the Taylors mostly lived apart – an arrangement which scandalised society and resulted in social isolation for the couple. It seems that John and Harriet worked well together, each inspiring the other, sparking ideas off each other.

Indeed, Mill attributes to Harriet a chapter 'On the Probable Future of the Labouring Classes' in his *The Principles of Political Economy* (1848) in which she argues for the vital importance of education for all in the future of the nation, both economically and socially. Her essay, *The Enfranchisement of Women* (1851), was actually published under John Mill's name. She believed strongly that women have access to the same jobs as men, and that they should not be forced to live in 'separate spheres' – views more radical even than those of Mill himself.

Harriet's husband, John Taylor, died in 1849, enabling her and Mill to be married at last in 1851. That autumn, the couple relocated to France to help with Harriet's tuberculosis. She died of respiratory disease in Avignon on 3 November 1858. John Stuart Mill's most famous work, *On Liberty*, which they had written as a team, was published in 1859 and was dedicated to Harriet.

Feminists

Many of the women featured in this book could be described as feminists to some extent and in one form or another of what is a particularly elusive term.

Mary Wollstonecraft (1759–1797)

Mary Wollstonecraft is a renowned English writer and philosopher best known for her advocacy of women's rights. Until very recently, in the late twentieth century, Wollstonecraft's unconventional life, which took in several irregular (as judged at the time) personal relationships, received more tabloidesque attention than her writing. Nowadays Wollstonecraft is regarded as one of the founding feminist philosophers, and feminists cite both her life and her works as important influences.

During her short life, she turned out novels, a travel book, a history of the French Revolution, a conduct book, and a book for children. However, Wollstonecraft is best known for *A Vindication of the Rights of Woman* (1792), in which she argues against the prevailing male notion that women are naturally inferior to men; it just looks that way because they did not benefit from a rigorous and extended education.

After two ill-starred affairs, with Henry Fuseli the artist, and Gilbert Imlay, an American businessman, author and diplomat by whom she had a daughter, Fanny Imlay, Wollstonecraft married the philosopher William Godwin, one of the architects of the anarchist movement.

Suicidal thoughts

Wollstonecraft was no stranger to the subject of suicidal thoughts, both in life and in her fiction. We know of two suicide attempts following the despair and rejection resulting from her fractured relationship with Gilbert Imlay. Both times she jumped into the Thames, but was rescued. These experiences, and her description of the subject in subsequent writings (for example in *Mary* and *The Wrongs of Woman*) informed how she equated suicide as an expression of protest against the oppression of women, particularly in marriage. She drew wives as being in 'marital enslavement' who saw suicide as a logical response to this tyranny. Her work also provides insights into the societal constraints faced by contemporary women and the concept of rational suicide in the Romantic era.[1]

Wollstonecraft died at the age of 38, leaving behind several unfinished manuscripts. She died of puerperal fever eleven days after giving birth to her second daughter. Godwin was left to raise Mary, along with her older half-sister, Fanny Imlay.

Her subsequent education was made up of her father's tutoring in a broad and diverse range of subjects, educational outings, access to his library and to the many intellectuals who called in on him, including Coleridge and the former vice-president of the United States Aaron Burr. There was also a governess and a tutor and she enjoyed access to many of her father's children's books on Roman and Greek history in manuscript.

Widower Godwin naively published *Memoirs of the Author of A Vindication of the Rights of Woman* (1798), exposing his former wife's bohemian, unorthodox lifestyle, which had the effect of destroying her reputation for nearly a century. However, with the emergence of feminism at the turn of the twentieth century, Mary Wollstonecraft's belief in women's equality and assessments of conventional femininity started to receive the attention and appraisal they deserve.

Note

1. Michelle Faubert, 'The Fictional Suicides of Mary Wollstonecraft', *Literature Compass*, 12, Special issue on Romanticism and Suicide, December 2015, 652–659.

Mary Shelley (1797–1851)

Mary Shelley, is, of course, the celebrated author of the classic Gothic novel *Frankenstein or, The Modern Prometheus* (1818), which is also remembered as one of our earliest examples of science fiction. She was the daughter of Mary Wollstonecraft and William Godwin, and wife of Percy Bysshe Shelley. She also had a hand in editing and promoting some of her husband's works.

The tempestuous life led by this gifted and inspirational woman is profoundly complicated and deeply tragic – not altogether surprising when you consider her similarly gifted parents and husband and the company she kept, which include Lord Byron, as well as the death of the Shelleys' premature daughter in infancy and the early death too of her husband.

As noted above, Mary's mother, Mary Wollstonecraft, died eleven days after giving birth to her. She was then raised by her father, who gave her that eclectic, informal education, inculcating in her his own anarchist politics and philosophical theories. When she was 4, her father married a neighbour, Mary Jane Clairmont; the two Marys never really got on.

In 1814, aged 17, Mary started a relationship with one of her father's acolytes, Percy Bysshe Shelley, who was already married to Harriet Westbrook. Mary and Percy began a series of clandestine meetings at her mother Mary Wollstonecraft's grave at St Pancras Old Church, and they fell in love – she was 16, and he was 21. Around 26 June 1814, Godwin lost her virginity to Shelley, which, according to tradition, happened in the churchyard.

Notwithstanding, accompanied by her stepsister, Claire Clairmont, she and Percy left for France as part of a tour of Europe. By the time they were back in England, the two were broke and Mary was pregnant with Percy's child; a further complication was that Shelley was having an affair with Claire Clairmont – a cause of some justified jealousy, even though Mary was an advocate of free love. The next two years were very difficult: she and Percy were socially ostracised, they were forever in debt and their prematurely born daughter died. This loss resulted in Mary Godwin's acute depression; she was haunted by visions of the baby but she conceived again and had recovered by the summer. They married in late 1816, after the suicide of Shelley's first wife, Harriet Westbrook, by drowning herself in the Serpentine.

Shelley and Harriet (aged 19 and 16) had eloped to Edinburgh in 1811 where they married, subsisting on borrowed money; their respective fathers had cut off their allowances. They were accompanied by Thomas Jefferson Hogg, a former fellow student at Oxford who became a close friend of Shelley's and an influence.

Shelley and Harriet stayed in Edinburgh for a month, with Hogg living with them. The three left for York in October, then Shelley went down to Sussex to iron out matters financial with his father, leaving Harriet behind with Hogg. Harriet revealed that Hogg had tried to seduce her in Shelley's absence.

In 1816, Percy, Mary and Claire Clairmont spent a summer with Lord Byron near Geneva where Mary conceived the idea for *Frankenstein* and Claire conceived a baby by Lord Byron. On returning to England that September, Mary and Percy moved – with Claire Clairmont, who took lodgings nearby – to Bath, intending to keep Claire's pregnancy secret. At Cologny, Mary Godwin had received two letters from her half-sister, Fanny Imlay, who alluded to her 'unhappy life'; on 9 October, Fanny wrote an 'alarming letter' from Bristol that sent Percy Shelley hot-foot to find her, but without success. On the morning of 10 October, Fanny Imlay was found dead in a room at a Swansea tavern, along with a suicide note and a bottle of laudanum. On 10 December, Percy Shelley's wife, Harriet, was, as noted, discovered drowned in the Serpentine.

Both suicides were hushed up and Harriet's family obstructed Percy Shelley's efforts – with the backing of Mary Godwin – to assume custody of his two children by Harriet. His lawyers advised him that to improve his case he might consider marrying so he and Mary, who was pregnant again, married on 30 December 1816 at St Mildred's Church, Bread Street, London.

The threat of debtors' prison hanging over their heads made the couple's decision to leave England for Italy on 12 March 1818 easy, taking Claire Clairmont and her daughter Alba with them. It seems they had no intention of returning. Their second and third children died – Clara, in September 1818 in Venice, and William, in June 1819 in Rome of malaria. Depression followed but Percy Florence Shelley was born; he was to be Mary's last and only surviving child.

In 1822, her husband tragically drowned when his boat sank during a storm near Viareggio. The next year, Mary Shelley returned to England to focus on raising her son and on her career as an author.

Mrs Shelley continued to put into practice her mother's feminist principles by extending aid to women scorned by society; she gave financial help to Mary Diana Dods, a single mother, illegitimate herself, and a lesbian; she charitably gave her the new identity of Walter Sholto Douglas (1790–1830), a Scottish author of books and stories. Despite being assigned a female identity, Douglas lived as a man, with most of his works published under the pseudonym David Lyndsay, and as the husband of Isabella Robinson.

Mary Shelley's last ten years were plagued by illness, probably caused by the brain tumour which killed her at the age of 53.

As with her mother, there was considerable salacious interest in the scandal surrounding Mary Shelley's life, and in *Frankenstein*, but in the 1970s the focus started to move on to a more scholarly examination of her other literary achievements, particularly her novels, which include *Valperga* (1823) and *Perkin Warbeck* (1830), the apocalyptic novel *The Last Man* (1826) and her final two novels, *Lodore* (1835) and *Falkner* (1837), as well as the travel book *Rambles in Germany and Italy* (1844), all of which support the growing consensus that Mary Shelley remained politically radical all her life.

Interestingly, Mary Shelley's works often argued that the ways to reform society lay in compromise and sympathy, particularly as practised by women in the family setting. In effect this was a direct challenge to the Romantic ethos promoted by poet Percy Shelley and the Enlightenment theories articulated by her father, William Godwin.

According to Betty T. Bennett, author of *Mary Wollstonecraft Shelley: An Introduction* (1998): 'Scholars now consider Mary Shelley to be a major Romantic figure, significant for her literary achievement and her political voice as a woman and a liberal.'

Mary Gray (1854–1941)

Mary Gray was a socialist activist and feminist, best remembered for her achievement in founding one of the first Socialist Sunday schools. Gray got an early taste of poverty when she was 15; she found work as a domestic servant, and in 1876 married Willie Gray, a stonemason and trade unionist who, because of his trade union affiliations, found it difficult at times to get work.

In 1887, Mary Gray joined the Social Democratic Federation (SDF). During the London dock strike of 1892, she ran a soup kitchen; her contact with the children of the workers was the trigger for her to found one of the first Socialist Sunday schools with the objective of providing a general education, as well as informing them about socialism. The first class comprised her own two children and one other, but the movement grew rapidly, and more than 200 Socialist Sunday Schools had been set up around the country by 1914.

Sunday schools of the socialist variety were a small part of a much bigger educational phenomenon. The Sunday school movement was a huge player in early education for the poorer elements of society. Its profound impact on child education and its effect on the difficult journey to more enlightened social welfare generally and universal education is often underestimated. The numbers of young minds involved and impacted alone are staggering.

The Protestant Sunday school first appeared in the eighteenth century to provide an education for poor, and particularly working children. Over the years the society has published its own textbooks and founded nearly 4,000 Sunday schools.

By 1785, 250,000 children (about 25 per cent of the eligible population) went to Sunday school every week, including the interdenominational Stockport Sunday school which financed and constructed a school for 5,000 scholars in 1805.

By 1835, the Sunday School Society had handed out 91,915 spelling books, 24,232 New Testaments and 5,360 Bibles.

Gray was elected to the Battersea Board of Guardians in 1895, serving until 1901, and from 1896 until 1903 she was a member of the SDF's executive.

Writers

Our survey of the convoluted lives of the Wollstonecrafts and the Shelleys brings us neatly into a section on the numerous women writers who found fame, recognition and some appreciation during the Industrial Revolution: they literally represent an explosion of talent and experimentation which assured women a rightful place in our peerless literary heritage, allowing many to become professional writers to rival their male counterparts.

Some seventeenth- and eighteenth-century women writers, Jane Austen, George Eliot and the Brontës, for example, initially resorted to publishing anonymously or pseudonymously in order to swerve round the steadfast male belief that a women couldn't possibly write a book and if they did, no one would buy it.

What the male-run publishing houses-cum-printers could not see or declined to notice was that women were uniquely placed to satisfy a market need amongst women for themes born of and indicative of the female imagination, female emotional experiences, and the real-life challenges encountered by women living and working in a stubbornly patriarchal society. The Blue Stockings Society helped a lot in showing how women could think and write in a scholarly fashion, how women could argue and debate on academic and philosophical matters, how there was a different view or argument to the sometimes hoary old man's tired and traditional way of seeing things. Women too helped develop the industrial and social novel (Elizabeth Gaskell being a prime example), identifying areas of inequality, discrimination against the poor as well against women, social marginalisation and abuse in the workplace and at home. They were also able to use their writing to promote their case for democratic voting rights and equal rights as workers, to promulgate a form of proto-feminism and win for themselves greater female participation in the arts and in public life.

The Blue Stockings Society and Elizabeth Montagu (1718–1800)

Grays Court in York is the oldest continuously occupied house in the country; parts of it date back to 1080. Elizabeth Montagu (née Robinson) was born here: she founded the Blue Stockings Society where literary topics were to be discussed, but politics, gossip and card-playing were barred. Men's matters, then, were definitely not on the agenda.

The society is dealt with in detail in my *Bridgerton Revisited*.

Women's private members' clubs provided another opportunity for educated middle-class women to flex their intellectual muscle. They are also dealt with in my *Bridgerton Revisited*.

Here then are some of the perhaps lesser-known women authors trailblazing during the Industrial Revolution.

Hannah Robertson (1724–1800?)

Hannah Robertson was a Scottish autobiographer and teacher. She claimed that her father, named Swan, was a son of Charles II. Robertson had an unhappy marriage to a wealthy man who went bankrupt in 1756. She paid off his debts by running a successful pub in Aberdeen. In 1771, he froze to death in a storm.

Robertson moved to Edinburgh, York, London, Northampton, and finally Birmingham; in all these places she taught filigree, japanning, and other fashionable accomplishments for young ladies. Her major work, *The Young Ladies School of Arts*, went into many editions in the decades after its initial publication in 1766. The fourth edition gives a good idea of the contents and exemplifies the fashion for prolix titles and her manifold skills and expertise. This was its full title:

> The young ladies school of arts. Containing a great variety of practical receipts, in gum-flowers, filligree, japanning, Shell-Work, Gilding, Painting, Cosmetics, Jellies, Preserves, Cakes, Cordials, Creams, Jamms, Pickles, Candying, Made Wines, Clear Starching, &c together with directions for breeding canary birds, and breeding, nursing, and ordering of the silk-worm. Also a great many Curious Receipts, both useful and entertaining, never before published. By Mrs. Hannah Robertson. The fourth edition, with large additions.

Her 1791 work is hardly less verbose and gives us a mini-autobiography, focusing on her reduced circumstances:

> The life of Mrs. Robertson, (a tale of truth as well as of sorrow) who, though a grand-daughter of Charles II. has been reduced, by a Variety of Very Uncommon Events, From splendid Affluence to the greatest poverty. And, After Having Buried Nine Children, is Obliged, At the age of Sixty-Seven, To earn a scanty Maintenance for herself and two Orphan Grand-Children, By teaching Embroidery, Filligree, and the Art of making Artificial Flowers. Derby: John Drewry, 1791.

Her inclusion of 'making artificial flowers' chimed nicely with the contemporary fashion and avid pursuit of floriography, the language of flowers: gifts of flowers, plants, and bespoke floral arrangements were deployed to send coded messages to the recipient, allowing the sender to express feelings of love and affection which could not be properly vocalised in polite society. Armed with floral dictionaries, lovers often exchanged small 'talking bouquets' or 'word poesy', called nosegays or tussie-mussies, which could be worn or carried as a fashion accessory.

Shakespeare, Austen, Charlotte and Emily Brontë, Frances Hodgson Burnett and Steinbeck among others, used the language of flowers in their writings.

Robertson's aim was to feed her family, and to help women who needed another income stream to support their families.

Elizabeth Raffald (1733–1781)

Before Mrs Beeton there was Manchester's Elizabeth Raffald. Elizabeth Raffald (née Whitaker) was author, innovator and entrepreneur. Born and raised in Doncaster, Raffald went into domestic service for fifteen years, ending as the housekeeper to the Warburton baronets at Arley Hall, Cheshire. She resigned her position when she married John, the estate's head gardener.

The couple moved to Manchester where the shrewd Raffald enterprisingly opened a register office to introduce domestic workers to employers – a kind of hiring fair in an office, or an early recruitment agency; she also ran a cookery school and sold food from the premises. In 1769 she published her blockbuster cookery book *The Experienced English Housekeeper*, which contains the first recipe for a 'Bride Cake' – the modern wedding cake. She is also probably the inventor of the Eccles cake.

To go with her recruitment office in August 1772, she published *The Manchester Directory*, a listing of 1,505 traders in Manchester – the first such listing for the up-and-coming town and one of the early forerunners of the business and trade directories which appeared at the end of the nineteenth century and to the end of the twentieth.

The Raffalds went on to run two post houses in Manchester and Salford before being beset by financial problems, not helped by John's heavy drinking. Elizabeth died suddenly in 1781 in mid-flow: she had just published the third edition of her directory and was still updating the eighth edition of her cookery book. After her death there were fifteen official editions of her cookery book, and twenty-three pirated ones (*plus ça change*). Her recipes were ruthlessly plagiarised.

Jane Marcet (1769–1858)

Marcet was an English salonnière and a groundbreaking writer of popular, explanatory, 'made simple'-type science books. She also showed her innovative publishing nous with her *Conversations on Political Economy* (1816), which explain the ideas of Adam Smith, Malthus and David Ricardo.

Jane Marcet was born in London on New Year's Day 1769, one of twelve children. She was home educated with her brothers on a diet including Latin, chemistry, biology and history, as well as those genteel subjects more usual for young ladies of the time.

She took over the running of the family and house at age 15, after her mother's death, helping to bring up her younger siblings. She was also enlisted as her widowed father's hostess, helping at the many parties of invited scientific and literary guests. In addition to all that, she cultivated an interest in painting on a visit to Italy with her father in 1796, and studied with Joshua Reynolds. This artistic training later enabled her to illustrate her own books.

One day, after proofing of one of her husband's books, Marcet decided she could write her own and started turning out expository books on chemistry, botany, religion and economics under the general series title 'Conversations'. In her prefaces, Marcet addressed the thorny subject of the day relating to whether such knowledge was suitable for women, arguing against the largely negative male objections and emphasising that public opinion supported her view.

The first was published in 1819, as *Conversations on Natural Philosophy* covering the basics of contemporary scientific knowledge: physics, mechanics, astronomy, the properties of fluids, air and optics. This established a template for her works: an imaginary dialogue between two pupils, Caroline and Emily, and their teacher, Mrs Bryant. Mrs Bryant studiously encouraged the girls to question and examine their ideas. The scientific content and the discursive process of sharing scientific knowledge were central to Marcet's content and her readers.

Marcet's next book, *Conversations on Chemistry, Intended More Especially for the Female Sex*, was published anonymously and became her best-known book on Humphry Davy, whose lectures she had attended. It was one of the first elementary science English language textbooks, complete with Marcet's drawings of chemical apparatus and emphasising the need for experiment and for theoretical rigour. Jane Marcet was identified as the author with the twelfth edition of 1832. The book went into sixteen editions in England, giving inspiration to a young Michael Faraday, who later claimed it had been influential in his own career.

Marcet's major contribution to elementary science and economics publishing was truly trailblazing, presenting the subjects for ordinary men and women in an accessible, uncomplicated and informal way. Girls and women benefited, having, until now, been denied access to such pedagogy in many institutions such as women's seminaries.

Mary Somerville said of Marcet: 'No one at this time can duly estimate the importance of Mrs Marcet's scientific works.'

Frances Milton Trollope (1779–1863)

Frances (Fanny) Trollope was a prolific and polemical writer. In 1809 she married Thomas Trollope, with whom she had seven children. In 1827, she emigrated to America with the aim of improving the family's financial circumstances; she didn't. When her hapless barrister husband went bankrupt and could no longer support the family, Frances, now 52, tried her hand at writing in order to bring in the bacon. She wrote forty books: six travelogues, thirty-four novels, numerous polemics and poems.

On returning to England, she published her first and bestselling travelogue, *Domestic Manners of the Americans* (1832), replete with waspish remarks on American society. Trollope criticised the rough manners of American men, and expressed a concern for American women who, she believed, were subjected to an oppressive evangelist religion that led to slavish devotion, mental confusion, and even illness. In short, 'bound in the iron chains of a most tyrannical fanaticism'.

'At one time or another during her life' – as Pamela Neville-Sington writes – 'her popularity matched, and even exceeded, that of Jane Austen, Charles Dickens, and her son Anthony.' Some of Fanny Trollope's novels contain profound social concern and demand social reform.

The Vicar of Wrexhill (1837), regarded as Trollope's best novel, is a stinging satire on Evangelicalism, which she despised.

She may well have been the first person to dare to suggest that women could be victims of sexually abusive priests. The education of women was Trollope's solution to the 'Woman Question', believing as she did that better-educated women would make better mothers and daughters.

Michael Armstrong, the Factory Boy (1840) is one of our first industrial novels. Andrzej Diniejko, writing in the Victorian Web,[1] tells us that:

> Trollope's maternal feminism is focused on the plight of child labourers. The novel, which addresses labour conditions in Manchester, exposes the abuses of child employment in early Victorian England. Trollope was probably inspired by *A Memoir of Robert Blincoe* (1828), a classic document about child labour and factory system, written by John Brown. She also made use of Parliamentary Blue Books, the *Sadler Report* (1832), and her own personal observations from her visits to factories in Manchester and Bradford.

Diniejko continues, explaining how the novel traces the outcome of 'miscalculated benevolence'. Sir Matthew Dowling, a wealthy industrialist, adopts an orphaned factory child called Michael Armstrong and takes him to his household. However, Michael feels isolated and abused in the house of his benefactor, who is soon bored with the boy and decides to send him away to the oppressive Deep Valley Mills where children are subjected to exhausting hard labour, starved and left prone to disease.

Fortunately, Michael escapes from the mill. The novel ends happily, with two unlikely cross-class marriages. Michael makes contact with his crippled brother Edward and his childhood friend, Fanny Fletcher – both were adopted by a wealthy and benevolent cotton heiress, Mary Brotherton, who is appalled by the brutality that was child labour. Michael marries Fanny, and Edward marries Mary, and they all emigrate to Germany so as to leave behind their traumatic experiences in England.

Trollope contends that individual philanthropy cannot solve the problems of mass poverty; her novel was influential in the revision of the Factories Act 1844. Trollope anticipated Disraeli and Gaskell in the vivid description of slum housing and appalling factory conditions, the intention being to prise open the eyes of her middle-class readers.[2]

Jessie Phillips, A Tale of the Present Day (1843) deals with the vexatious issues of seduction and infanticide and the inadequacy of the New Poor Law 1834, revealing the sorry plight of poor 'fallen' women who were unable to bring charges against fathers of their illegitimate children on account of the 1834 Bastardy Act. Other main themes include needleworking, workhouse living, suicide and death. The government enacted legislation which held fathers financially responsible for their illegitimate children.

Like Gaskell, Trollope, as noted, exhorted women to get themselves a better education in order to become self-sufficient. She also urged middle-class women to articulate their social concerns in the public sphere, not allowing domesticity and motherhood to totally preoccupy and constrain them.

The crucial legacy of Trollope's novels was that they aroused the public conscience and provoked social reforms that protected victimised women and children and held fathers to account. They were highly influential in influencing women's writing that came after her.

Notes

1. 'Frances Trollope: a Maternal Feminist and Social Reformer', Dr Andrzej Diniejko, Senior Lecturer in English Literature and Culture, Warsaw University; Contributing Editor, Poland. https://victorianweb.org/authors/francestrollope/diniejko.html
2. For details see Paul Chrystal, *Factory Girls: The Working Lives of Women and Children*, Pen & Sword, Barnsley, 2022.

Jane Taylor (1783–1824)

Jane Taylor was an English poet and novelist best known for the lyrics of 'Twinkle, Twinkle, Little Star'. Her first work was as contributor to the anthology *Original Poems for Infant Minds by Several Young Persons*, published anonymously. This was followed by *Rhymes for the Nursery* in 1806 and *Hymns for Infant Minds* in 1810 with her sister Ann Taylor. The standout piece in the 1806 collection is 'The Star', known now as 'Twinkle, Twinkle, Little Star'.

It is now sung to the tune of the French melody 'Ah! vous dirai-je, maman', released in 1761 and later arranged by several composers, including Mozart with Twelve Variations on 'Ah vous dirai-je, Maman'. The English lyrics have five stanzas, although only the first is widely known.

Jane's *Essays in Rhyme* was published in 1816, and 'contained some significant poetry'. In the fictional *Correspondence between a Mother and Her Daughter at School* (1817), Taylor collaborated with her mother. *The Family Mansion. A Tale* appeared in 1819, and *Practical Hints to Young Females* before 1822.

Ann's poem 'The Maniac's Song', published in the *Associate Minstrels* (1810), was probably the finest short poem by either sister, and it has been postulated as an inspiration for Keats's 'La Belle Dame sans Merci'.[1]

The best known of many parodies of 'Twinkle, Twinkle, Little Star' is a poem recited by the Mad Hatter in Lewis Carroll's *Alice's Adventures in Wonderland* (1865).

Note

1. Lynette Felber, 'Ann Taylor's "The Maniac's Song": an unacknowledged source for Keats's "La Belle Dame sans Merci"', *ANQ: A Quarterly Journal of Short Articles, Notes, and Reviews*, 17(1), 2004.

Maria Jane Jewsbury (1800–1833)

Maria Jane Jewsbury was a writer, poet and reviewer who wrote for the *Manchester Gazette* and *The Athenaeum* – all fitted in while bringing up her younger siblings for twelve years.

Jewsbury's father was the master of a cotton factory, but as a consequence of the War of 1812 with America, which impacted on the cotton business, the family had to move to George Street in 1818, when the business failed. Jewsbury's mother died one month after giving birth. It was then that aged 19, Jewsbury took on the mother's role for the household, so that her father could continue working.

Her *Phantasmagoria* or, *Sketches of life and literature* (1825), *The Three Histories* (1830) and *Letters to the Young* (1837) were her most popular works and are the most enduring.

Phantasmagoria takes readers on a thrilling journey of literature and the human experience. The author unearths the historical evolution of writing and reading, contrasting eras where literacy was a rarity with the explosion of books in the nineteenth century. The work critically examines the trends and pitfalls of contemporary literature, particularly the rise of 'religious novels' and their impact on readers.

Her literary circle of friends included William Wordsworth and Dorothy, Sara Coleridge and Thomas De Quincey.

In 1832, she married Rev. William Kew Fletcher (d.1867); the couple sailed for India where she kept a journal and had poetry printed in *The Athenaeum* as 'The Oceanides'.

Jewsbury fell ill in June 1833 and died of cholera at Poona.

The Three Histories (1830) is considered her best work. The histories in question are those of an Enthusiast, a Nonchalant, and a Realist.

Elizabeth Gaskell (1810–1865)

Gaskell was well-schooled in the reality of squalor and poverty through the research she did for her social novel, *Mary Barton* (1848) investigating in great detail the plight of the Manchester poor between 1839 and 1842. Although she was born in the leafy comfort of middle-class Cheyne Walk (number 93) in Chelsea, she was well qualified to expatiate on the disparities between north and south living by the time she reached Manchester via Warwickshire and Knutsford.

When, in 1850, the Gaskells moved to Manchester, first to middle-class Dover Street off Oxford Road, and then Plymouth Grove, Elizabeth, staying rooted, brought her cow with her and would gladly walk up to three miles to help another person in distress. A lottery grant was used to help restore the house's garden, where the Gaskells once kept pigs, poultry and the cow.

In Manchester, Elizabeth wrote her remaining literary works largely in the industrial genre. Gaskell documented Manchester's industrial and social revolution from her kitchen desk after the family moved to the house in 1850. Her Milton in *North and South* is our Manchester.

Indeed, Elizabeth felt some guilt living in their well-appointed villa while many others close by subsisted in insanitary hovels. Nevertheless, the Gaskells were never what you might call frugal, what with their £150 yearly rent, twenty rooms and a cook, several maids, an odd-job man for outdoor work, as well as a washerwoman and a seamstress.

Isabella Banks (1821–1897), aka Mrs G. Linnaeus Banks

Isabella Banks was a novelist and poet mainly remembered today for her classic book *The Manchester Man*, published in 1876.

Things started to take off for Isabella when *The Manchester Guardian* published her poem, 'A Dying Girl to her Mother', in 1837. Around the same time she was frequenting the Sun Inn on Long Millgate in Manchester, a pub popular with poets, authors, and other working-class intellectuals who became a poetry collective known as the Sun Inn Group.

Her output includes *Bond Slaves – The Story of Struggle* (1893), a social novel about Luddites in the north of England. But it is *The Manchester Man* for which Isabella Banks is best remembered – first serialised in *Cassell's Magazine* between January and November 1874, before being published in three volumes in 1876.

It remains an important social and historical novel, following the rise of Jabez Clegg, the eponymous 'Manchester Man', from the Napoleonic Wars to the First Reform Act. His personal fortunes create a tale of romance and melodrama from apprentice to master and from poverty to wealth, reflecting the growth and prosperity of Manchester.

The Manchester Man is remarkable for its historical detail.[1] It contains vivid accounts of early nineteenth-century textiles and fashions, historical characters, and historical events including *Peterloo* and the sinking of the *Emma*,[2] based on unpublished eyewitness accounts. The novel includes appendices in which the author delineates the historical versus fictional content.

The book is still read throughout the world and remains in print. A quotation from the novel forms the epitaph on the Southern Cemetery tombstone of Tony Wilson, one of the founders of Factory Records in Manchester:[3]

> Mutability is the epitaph of worlds / Change alone is changeless / People drop out of the history of a life as of a land though their work or their influence remains.

Notes

1. Barbara Brill, 'Some Manchester Chroniclers', *Library Review*, 24(4), pp. 147–153, 1973.
2. https://en.wikipedia.org/wiki/Emma_(1828_ship)
3. Patrick Burgoyne (22 October 2010), 'Saville and Kelly's memorial to Tony Wilson', www.anothermag.com/art-photography/2651/peter-saville-on-tony-wilsons-headstone

Ellen Johnston (1835–1874) – 'the factory girl'

Ellen Johnston epitomised the idea of 'the factory girl' of nineteenth-century Scotland. When Ellen was only seven months old, her father emigrated to America to follow his dreams; Mary Bisland, her mother, and Ellen went to live with Mary's father, while Mary took work as a seamstress and milliner. In 1844, believing that her husband was dead, Mary remarried and with Ellen (now 8) moved in to live with Ellen's stepfather. When he found out years later, Ellen's father, James, killed himself.

Ellen herself tells us quite confidently in her journal that she spent '9 months at school when I could read the English Language and Scottish dialect with almost any classic scholar'. Alison Chapman, in her 'The Scottish Women Poets' website, reveals that 'It is presumed that Ellen was sexually abused by her step-father, who also had her working in the factories from a young age as a power-loom weaver' to cure her of her bookish pretensions.

In 1852, Ellen gave birth to her illegitimate daughter, Mary Achenvole. In 1854, the *Glasgow Examiner* published Ellen's poem 'Lord Raglan's Address to the Allied Armies'. In 1863, Ellen was dismissed from the Verdant Factory and ill-advisedly sued the factory, resulting in her blacklisting amongst other perspective employers. The lawsuit left Ellen walking the streets of Dundee as a 'famished and persecuted factory exile' until April 1864; Ellen recalled in her journals that 'even my poor ignorant deluded sister went so far as to assault me on the streets, spit in my face, and even several times dragged the skirts from my dress'.

Most of Ellen's work was published in periodicals and newspapers under the pseudonym of the 'Factory Girl'; she benefited from the repeal of the 'Stamp Act' which allowed for more circulation of working-class magazines. In 1867, Ellen published her *Autobiographies, Poems, Songs, and the 'Factory Girl'* by subscription in Glasgow by William Love: 800 copies were sold. Ellen's inscription read: 'to all men and women of every class, sect, and party, who by their skill, labour, science, art, literature, and poetry, promote the moral and social elevation of humanity.'

Ellen gave up factory work; March 1869 saw a second edition.

Alison Chapman concludes that Ellen Johnston defined herself as a Scots woman with 'highly susceptible and sympathetic natures – physical, intellectual, and moral'. She was greatly inspired by the novels of Sir Walter Scott (even though he was no fan of haughty heroines and believed, as in *Ivanhoe*, that men should conduct themselves as mediaeval knights did). Ellen Johnston is rightly remembered because she personified the working-class factory girl.

In 1991, her poem written in dialect, 'The Last Sark', was published in *An Anthology of Scottish Women Poets*.

Charlotte Brown Carmichael Stopes (1840–1929)

Edinburgh-born Charlotte was also known as C. C. Stopes – academic, author, and campaigner for women's rights. At age 21 she published a collection, *Alice Errol and Other Tales*, and went on to take posts as a governess, one of the few careers available to educated young women in the 1860s and early 1870s.

Charlotte Carmichael had become a member by 1866 of Sarah Mair's Ladies' Edinburgh Debating Society, which published *The Attempt*; Charlotte published a number of articles therein. In 1867, Mary Crudelius presented her idea of convening classes for women at a university level under the auspices of the Edinburgh Ladies' Educational Association. Carmichael pledged her willingness to attend and bought along another twelve interested women. The first classes began in 1868, taught by Professor David Masson, Professor of English Literature at Edinburgh University, 'at a time when the University was not open to women and courses were given to them privately by the male Professors'.

Although women were barred from taking degrees, she achieved the highest certificate then available to a female student, in subjects as diverse as literature, philosophy and science, achieving first-class honours. She 'was the first woman in Scotland to gain a Certificate of Arts'. Carmichael used her education to assist the advancement of women and pursued her own scholarly interests in the English Renaissance, particularly Shakespearean literary history.

She authored several books on the life and work of Shakespeare but her most successful publication was *British Freewomen: Their Historical Privilege* (1894), a book which influenced and inspired the early twentieth-century British women's suffrage movement. It ran to several editions and was a key text for the British female suffrage movement. Helen Blackburn, who had supplied Stopes with notes from her own research to help the enterprise, purchased the whole first edition, many copies of which were sent to members of the House of Commons. Laura E. Nym Mayhall notes that *British Freewomen* was 'perhaps the single most influential text in casting women's struggle for the vote'; Stopes's arguments were frequently cited by 'suffragists of all stripes in making the case for women's suffrage in print, before crowds, and in the courtroom'.

She also had a passionate interest in Victorian dress reform and the need for comfortable clothes for women. She was a member of the Rational Dress Society. At a meeting of the British Association for the Advancement of Science in Newcastle-upon-Tyne in 1889, she stunned the proceedings by organising an impromptu session where she introduced sensible dress to a wide audience.

She married Henry Stopes, a palaeontologist, brewer and engineer. They had two daughters, the eldest being Marie Stopes, famous for her pioneering work in birth control.

Mary Augusta Ward (1851–1920)

Mary Augusta Ward wrote under her married name as Mrs Humphry Ward. Her aim in life was to improve education for the poor and to that end she established a settlement in London. Unlike many of her contemporary trailblazers, she staunchly opposed giving women the vote, and in 1908 she became the founding President of the Women's National Anti-Suffrage League.

Mary Augusta was born in Hobart, Tasmania into an intellectual family of writers and educationalists who included the Huxleys and Arnolds in their number. At 16 she returned to live with her parents at Oxford, where her father lectured in history.

On 6 April 1872, not yet 21 years old, Mary married Humphry Ward, a fellow and tutor of Brasenose College, and also a writer and editor. She was by now conversant in French, German, Italian, Latin and Greek and was developing a keen interest in social and educational service as well as dabbling in literature. She mastered Spanish and in 1877 was writing many of the lives of early Spanish ecclesiastics for the *Dictionary of Christian Biography* edited by Dr William Smith and Dr Henry Wace.

Ward was all for opening up Oxford University to female students and joined the Lectures for Women Committee, which met from 1873 and organised lectures with an optional final examination for women. With others she formed the Association for the Education of Women, which supported the opening of halls for women students in Oxford.

Mary Ward began writing articles for *Macmillan's Magazine* while writing a children's book published in 1881 as *Milly and Olly*, followed in 1884 by a more ambitious study of modern life, *Miss Bretherton*, the story of an actress. Her novels were imbued with strong religious content consonant with the Victorian values she herself practised. Her popularity spread to the United States where *Lady Rose's Daughter* was the bestselling novel in 1903, as was *The Marriage of William Ashe* in 1905. Ward's most popular novel was easily the religious 'novel with a purpose' *Robert Elsmere*.

As noted, Ward worked and taught among the poor and as an educator in the residential settlement movements she founded; her declared aim was 'equalisation' in society, and she established educational settlements first at Marchmont Hall and later at what is today called Mary Ward House on Tavistock Place in Bloomsbury. It is now the Mary Ward Centre and continues as an adult education college; affiliated with it is the Mary Ward Legal Centre.

Ward was also a vigorous campaigner against women getting the vote, starting up and editing the *Anti-Suffrage Review* in which she published many articles on the subject. She projected her beliefs in two of her novels, *The Testing of Diana Mallory* and *Delia Blanchflower*, to criticise the suffragettes. In a 1909 piece in *The Times*, Ward, untypical of other trailblazers, wrote that constitutional, legal, financial, military, and international problems were problems only men could solve. Despite this, she did encourage women having a voice in local government and other rights that were anathema to the men's anti-suffrage movement. Julia Stephen, Virginia Woolf's mother, recommended Florence Nightingale, Octavia Hill and Ward as excellent role models for her daughters.

* * *

Other prominent writers of the day of course include: Jane Austen (1775–1815), Anne, Charlotte and Emily Brontë; Elizabeth Barrett Browning (1806–1861); George Eliot (1819–1880); Christina Rossetti (1830–1894); and Francis Hodgson Burnett (1849–1924).

Other less well-known writers include: Ellen Price/Mrs Henry Wood (1814–1887) and Rhoda Broughton (1840–1920). Broughton was known as 'a queen of the circulating libraries'. Her *Dear Faustina* (1897) is remembered for its homoeroticism while *Lavinia* (1902) depicts a seemingly 'unmanly' young man, who would have much preferred to have been born as a woman. Then there was Sydney, Lady Morgan (c.1778–1859), the Irish novelist, best known for *The Wild Irish Girl* (1806), a romantic, 'proto-feminist' novel imbued with political and patriotic overtones. Her work, including continental travelogues, sparked controversy and suffered censorship.

Artists

Anna Maria Garthwaite (1688–1763)

Anna Maria Garthwaite is one of the pre-eminent English textile designers of the mid-eighteenth century: her vividly water-coloured floral designs and the resulting silks have been admired the world over. She moved from Grantham to York with her twice-widowed sister Mary from 1726 to 1728 and then removed to Princes Street (now Princelet Street) in the silk-weaving district of Spitalfields, London. There Anna Maria created over 1,000 designs for woven silks over the next thirty years. Nearly 900 of her original designs from the 1720s to 1756 have survived and now reside in the Victoria and Albert Museum – many annotated with weaving instructions and the names of the weavers to whom they were sold.

Garthwaite has been called the 'pre-eminent silk designer of her period'.[1] Malachy Postlethwayt (c. 1707–1767), in *The Universal Dictionary of Trade and Commerce* of 1751, listed Garthwaite as one of three designers who had 'introduced the Principles of Painting into the loom'.[2]

Notes

1. Moira Thunder, 'Improving Design for Woven Silks', *Journal of Design History* 17(1), 2004, pp. 5–27.
2. Albany Institute of History and Art. Archived 13 May 2008 at the Wayback Machine.

Mary Delany, aka Mary Pendarves (1700–1788)

As with a number of her contemporaries, widowhood opened up new vistas and new opportunities for Mrs Pendarves. She found that widows, unlike unmarried women, were free to move as they wished in society, and for the first time in her life, Mrs Pendarves could pursue her own interests without being bothered by any man. She was clearly dissatisfied with the options available to women of her day. She wrote:

> Why must women be *driven to the necessity* of marrying? a state that should always be a matter of *choice*! and if a young woman has not fortune sufficient to maintain her in the situation she has been bred to, what can she do, but marry?[1]

In 1769, she translated William Hudson's (1762) *Flora Anglica* into English. Although her translation went unpublished, Delany's text provides a clear glimpse into her botanical knowledge.[2]

Mary Delany left it very late in her life to flourish and blossom. The Blue Stockings Society veteran was 72 when she took to producing her 985 exquisite, extraordinarily detailed floral collages, causing her to exclaim: *I have invented a new way of imitating flowers!* And indeed she had. It was at her friend the Duchess of Portland's house that Delany first used paper to model botanically accurate depictions of flowers.

In 1771, by now a widow in her early 70s, Mary took up decoupage, all the rage with fashion ladies of the court. The results were detailed and botanically accurate depictions of plants on tissue paper coloured by hand. She created 985 of these beautiful works, her 'Paper Mosaiks [sic]',[3] from the age of 71 to 88, when her eyesight failed her.[4]

> With the plant specimen set before her she cut minute particles of coloured paper to represent the petals, stamens, calyx, leaves, veins, stalk and other parts of the plant, and, using lighter and darker paper to form the shading, she stuck them on a black background. By placing one piece of paper upon another she sometimes built up several layers and in a complete picture there might be hundreds of pieces to form one plant. It is thought she first dissected each plant so that she might examine it carefully for accurate portrayal…[5]

The British Museum describes how

> Botanist Joseph Banks sent flowers from the Royal Botanic Gardens at Kew for Mary to copy: her pieces were so accurate that he described the collages as the only imitations of nature he had seen from which one could 'describe botanically any plant without the least fear of committing an error'.[6]

Mary had known many of the bright lights of her day and had corresponded with Jonathan Swift and Sir Joseph Banks; most importantly, she left a detailed picture of polite English society of the eighteenth century in her six volumes of *Autobiography and Letters* (ed. Lady Llanover, 1861–1862).

Notes

1. Ruth Hayden, *Mrs Delany: Her Life and Her Flowers*, London, 1980.
2. T. Cardoso (2023), 'Mary Delany's *British Flora* (1769): Female Agency in the Translation of Science'. In K. Bennett & R. M. Puga (Eds.), *Translation and Transposition in the Early Modern Period* (pp. 91–110), Routledge.
3. Mark Laird and Alicia Weisberg-Roberts (eds.), *Mrs. Delany and Her Circle*, New Haven, 2009.
4. Laird and Weisberg-Roberts, 2009, p. 36.
5. Hayden, 1980, pp. 132–133.
6. www.britishmuseum.org/blog/late-bloomer-exquisite-craft-mary-delany

Elizabeth Blackwell (1700–1758)

Records show that Elizabeth Blackwell studied as a midwife but gave it up due to 'the ignorance and low character of the women who at that time followed the same calling'.

Elizabeth Blackwell also made a significant contribution to the art of botanical illustration. Her multi-volume work, *A Curious Herbal*, published in the 1730s, was a cutting-edge, invaluable resource for doctors and apothecaries in the eighteenth century and later. It is one of the earliest botanical books to have been compiled by a woman and came about when Elizabeth had to find an income when her husband Alexander ended up in Highgate debtors' prison. Described as an 'ingenious lady', she set to work on an illustrated herbal which capitalised on her artistic and observational skills, and her interest in botany. She mainly researched and sketched plant specimens in the Chelsea Physic Garden.

Elizabeth will have been a publisher's dream author given that she did the sketching, engraving, and hand colouring for all of the illustrations, as well as getting personally involved in the business of publishing and marketing the book. The final work contained 500 illustrations which were published in 125 weekly instalments from 1736 to 1739. The creation of this landmark herbal took several years, and was a monumental endeavour. It was an instant success, and the profits helped save the family from ruin.

A Curious Herbal enjoyed Europe-wide fame. It met the need for an updated herbal: physicians and apothecaries praised the book while the Royal

College of Physicians endorsed it. A second edition was published around 1760, revised and enlarge by Dr Christoph Jacob Trew. Carl Linnaeus knew Elizabeth's work and gave her the title of 'Botanica Blackwellia'.

Mary Linwood (1755–1845)

Leicester's Mary Linwood was a celebrity artist in the early 1800s but, despite moving in a circle which included the rich and powerful, where she was highly respected, she has largely been forgotten. Contemporary textile artist Ruth Singer has been exploring the life and legacy of Linwood, rediscovering Linwood in effect.

> Mary Linwood made detailed embroidered versions of famous paintings by British artists in a technique called needle painting [or crewelwork]. Linwood was an innovator and entrepreneur as well as a talented artist. Alongside her Leicester school for young ladies, she exhibited her embroidered works in touring exhibitions and created the first gallery to be run by a woman in London.[1]

For nearly seventy-five years, she worked away in worsted embroidery to produce a portfolio of over 100 pictures, mostly full-size copies of old masters. These achieved great fame from the time of her first London exhibition in 1787.

The Greater Wigston Historical Society Bulletin 79 tells us that:

> Her mother opened a school in Belgrave Gate. When her mother died Mary took over the school and continued it for 50 years. She began creating pictures in needlework when aged 13, and was still doing so at 79. Many were copies of paintings by famous artists such as Gainsborough, Reynolds and Stubbs. In 1766 her work was shown at the Royal Academy in London. On one occasion her copy of a painting by the Italian artist Salvator Rosa (1615–1673) sold for more than the original. One of her own designs, the *Judgement of Cain*, took ten years to complete. Later she had her own exhibition which also went on tour. She met most of the crowned heads of Europe. She exhibited in Russia and Catherine the Great offered £40,000 for the whole collection while the Tsar offered her £3,000 for one example. However, Mary refused as she wished her work to remain in England. She received a medal in 1790 from

> the Society of Arts and was invited to an audience with Queen Charlotte … She never married and was the last person in Leicester to use a Sedan chair.[2]

Her exhibition in Leicester Square was a first because it was lit by gaslight and featured theatrical silver and gold curtaining.[3] Linwood's many subjects included Lady Jane Grey and Napoleon who conferred on her the Freedom of Paris in 1803.[4] Her annual shows attracted 40,000 customers a year, about the same as Madame Tussauds. The pictures continued to be exhibited in Leicester Square continuously until her death in 1845.

John Constable's (1776–1837) first commission was to paint in the background in one of her works.

Linwood refused an offer of 3,000 guineas for her version of Carlo Dolci's *Salvator Mundi*, and instead bequeathed it to Queen Victoria.

Notes

1. https://ruthsinger.com/mary-linwood. Needle painting and crewelwork are both hand embroidery techniques, but they differ in the materials used, stitches, and the overall aesthetic. Crewelwork uses wool yarn on fabric, often featuring bold, traditional designs, while needle painting, also known as thread painting, employs fine threads (like silk or cotton) and long and short stitches to create realistic, shaded images. The earliest surviving example of crewelwork is the Bayeux Tapestry, which is not really a tapestry at all.
2. Page 6.
3. Clare Hunter, *Threads of Life: A History of the World through the Eye of a Needle*, London, 2019.
4. 'Leicester Chronicler – Mary Linwood'.

Anna Atkins (1799–1871)

Botanical artist, collector and photographer Anna Atkins is regarded as the first person to illustrate a book with photographic images.

Kerry Lotzof of the Natural History Museum tells us that her cyanotypes used light exposure and a simple chemical process to produce highly detailed blueprints of botanical specimens.[1] Anna's innovative use of new photographic technologies represented a unique merging of art and science, and exemplified the huge potential of using photography in books.

Andrea Hart, Library Special Collections Manager at the Museum, adds:

> With the introduction of photography, you get a whole new opening up of how natural history and science can be presented in print.
> Before Atkins's book on British algae and the photographic process, botanical images would have been restricted to the traditional printing processes of engraving or woodcuts, although the art of nature printing was also in its early stages around Atkins's time.

Hart continues to reveal an important fact relating to women in science at the time:

> She lost her mother very early in life and as a result formed a close bond with her father [who] was a well-respected scientist and the first president of the Royal Entomological Society of London, which in turn opened doors for Anna to participate in science that would not have been possible for many women at that time.[2]

In general, women, as we have seen, were largely prevented from practising science professionally, preserving the stranglehold men had on the academic pursuit of the sciences. Botany, however, was a subject that was accessible to all – in particular botanical art and illustration, which were considered a suitably genteel pastime for women. Atkins was elected a member of the London Botanical Society in 1839.

Anna self-published her exceptionally detailed and highly meticulous images using the cyanotype photographic process in her 1843 book, *Photographs of British Algae: Cyanotype Impressions.*

Two more volumes followed between 1843 and 1853. In the volumes held at the Natural History Museum there are 411 plates each, with their scientific names handwritten by Anna. There are only twenty known extant copies, fifteen of which are substantially complete. She made every print herself.

In her early 20s, Atkins had made 256 scientifically accurate drawings of shells, which were published in her father's English translation of Jean Baptiste de Lamarck's catalogue *Genera of Shells (1822–24)*. Anna supplied specimens to leading botanists at Kew Gardens. Her superb herbarium of British plants was gifted to the British Museum in 1865.

Atkins also published five novels between 1852 and 1863: they included *The Perils of Fashion*, *Murder Will Out: A Story of Real Life*, and *A Page from the Peerage*. In the 1850s, Atkins collaborated with Anne Dixon (1799–1864), who was 'like a sister' to her, to produce at least three albums of cyanotype photograms:

- *Cyanotypes of British and Foreign Ferns* (1853), now in the J. Paul Getty Museum.
- *Cyanotypes of British and Foreign Flowering Plants and Ferns* (1854), disassembled pages of which are held by various museums and collectors.

Notes

1. www.nhm.ac.uk/discover/anna-atkins-cyanotypes-the-first-book-of-photographs.html
2. Her father, John George Children, was also an eminent chemist, mineralogist, and zoologist.

Mary Ellen Best (1809-1891)

Mary Ellen Best was a kitchen sink social history artist. Born in York, she produced most of her work in the 1830s, usually working in watercolour, usually painting scenes of domestic life depicting families at tables, in kitchens, and domestic staff. She painted a number of watercolours of her home in Castlegate, including interior portraits of the drawing room, dining room (1838), and one of herself working in her painting room. She also painted other houses including the drawing room of Naburn Hall. After marrying Johan Anton and moving to Nuremberg, Frankfurt, Worms, and Darmstadt, she continued to paint the houses they lived in.

Between 1828 and 1840, two-thirds of Best's portraits were commissioned by women. Caroline Davidson, Best's biographer, calculated that she produced more than 1,500 paintings in total.

Some of Best's drawings were handed down to her great-granddaughter, and forty-seven were sold in January 1983 in New York City; York Art Gallery obtained three of them. The paintings are of historical interest as an invaluable record of Yorkshire and European life in the 1830s and 1840s.

Constance Fox Talbot (1811–1880)

Constance Talbot was an English artist credited as the first woman ever to take a photograph – a somewhat hazy image of a short verse by the Irish poet Thomas Moore.

She married William Henry Fox Talbot, a key player in the development of photography in the 1830s and 1840s. Her watercolours and drawings remained locked away at Lacock Abbey, Fox Talbot's home, until they were digitised by the National Trust and made publicly available.

Louise Jopling (1843–1933)

Louise Jane Jopling was born in Manchester; her *Five O'Clock Tea* sold for £400 in 1874. Her *Five Sisters of York* (the resplendent window) was shown at the Philadelphia Exposition in 1876, her *The Modern Cinderella* at the Paris Exposition of 1878 and *It Might Have Been* at the opening exhibition at the Grosvenor Gallery in 1877. Stunning portraits of Jopling by Whistler in 1877 (now in Glasgow University's Hunterian Art Gallery) and by Millais in 1879 (now in the National Portrait Gallery) only added to her fame. She frequently appeared in, and was photographed for, the society pages of the day, and admired for her fashionable dress sense and social success: 'her studio parties are always interesting and she knows so many people who are always somebody in literature and art'.

Her social circle included James McNeill Whistler, Oscar Wilde, Kate Perugini (née Dickens) and Ellen Terry. Importantly she established her own school of painting for women and, also in 1887, wrote several articles on the teaching of art. She championed the right of female art students to work directly from live nude models as the stuffy Royal Academy only allowed its female students to 'observe' male models 'carefully draped' in 1893; her friend Whistler gave out the prizes at her school.

Louise Jopling supported the National Union of Women's Suffrage Societies, and was active in feminist causes. She was a vice-president of the Healthy and Artistic Dress Union, an organisation promoting dress reform during the 1890s and early 1900s.

In common with her contemporaries, she faced recurring rejection as a 'lady' artist, battling to succeed in an antediluvian masculine world where gender was more important than obvious talent. Her achievements are all the more impressive given that, as a woman, her work did not attract the bloated fees that male artists were raking in at the time.

In 1902, she became the first woman to be elected to the Royal Society of British Artists, RBA. She was a member of the Women's Tax Resistance

League – 'No Taxation without Representation'. Princess Sophia Duleep Singh, another eminent member, had a precious diamond ring of hers seized by bailiffs when she flatly refused to pay her servants' taxes or her dog licence. When the ring went up for sale at auction, fellow activists ensured that no one dared buy the ring. The successful bidder was none other than Louise Jopling, who immediately returned it to the princess.

She was also sponsor of the 'Suffrage Atelier', a cooperative which created propaganda for the women's suffrage movement to use in poster campaigns and marches and to sell at stalls.

Louise published a book of art instruction to help budding women artists and an autobiography, *Twenty Years of My Life*. She also wrote some poetry and journalism, contributed illustrations to *Punch* and inspired some of George du Maurier's cartoons satirising the fashions and manners of society.

Annie Swynnerton (1844–1933)

Annie Louisa Swynnerton, ARA, was born in Hulme, Manchester in 1844. She painted and sold watercolours to supplement the family's income; like her, sisters Julia and Emily also studied at Manchester School of Art and are recorded as being prize-winning students in *The Manchester Guardian* in 1873 (24 December, p. 8.) From 1874 to 1876 she studied in Rome with her friend, Susan Isabel Dacre; the women then attended the Académie Julian in Paris from 1877 to 1880. The catalogue for the Tate exhibition, 'Exposed: The Victorian Nude', states that 'Rome-based Annie Swynnerton was one of the most daring female painters of the nude, often shocking audiences with her robustly painted figures'.[1]

John Singer Sargent admired her work and with his help she became the first elected woman member at the Royal Academy of Arts in 1922. Swynnerton then was nothing if not well connected: she painted portraits of members of the Garrett family, including Agnes Garrett, Dame Millicent Garrett Fawcett, which hangs at the Tate Gallery, and Louisa Garrett Anderson. Annie was close to leading suffragists of the day, not least the Pankhursts. She also painted portraits of Henry James and the Rev. William Gaskell. Dacre and Swynnerton shared a studio and, in 1879, the pair founded the Manchester Society of Women Painters, which offered art education and exhibitions.

1893 saw *Florence Nightingale at Scutari* shown at the Women's Exhibition at the Chicago World's Exposition. Hilary Fraser, author of *Women Writing Art History in the Nineteenth Century*, tells us the work showed the manner with which women artists could convey tender feeling with strong artistic composition and colour. Swynnerton was included in the 2018 exhibit *Women in Paris 1850–1900.*

Note

1. *Exposed: The Victorian Nude*, Tate Publishing, 2001.

Elizabeth Southerden Thompson (1846–1933)

Elizabeth Thompson, later known as Lady Butler, specialised in depicting scenes from British military campaigns and battles, including the Crimean War and the Napoleonic Wars. The works she is best remembered for include *The Roll Call* (purchased by Queen Victoria), *The Defence of Rorke's Drift*, and *Scotland Forever!* (the Scots Greys at Waterloo). In her 1922 autobiography, she says: 'I never painted for the glory of war, but to portray its pathos and heroism.'

In 1866, she enrolled in the Female School of Art in South Kensington; the following year *Bavarian Artillery Going into Action* was shown at the prestigious Dudley Gallery in Piccadilly, one of the galleries preferred by women artists. The same year, she exhibited *Horses in Sunshine* at the Society of Female Artists. With her *Missing* (1873), a Franco-Prussian War battle scene showing the common soldiers' suffering and heroism, she won her first submission to the Royal Academy. Her *The Roll Call*, which showed a line of soldiers exhausted by conflict, was shown in 1874 at the Royal Academy Summer Exhibition and was so popular that a policeman was posted next to the painting in to keep the crowds in order. Butler later wrote that after the Summer Exhibition opened, she awoke to find herself famous.

Her fame only increased and, apparently, was boosted when people discovered that she was both young and attractive, qualities not usually associated with military art (!). The jingoism and upsurge in Victorian pride for the expanding British Empire did her no harm either.[1]

Note

1. Paul Usherwood and Jenny Spencer-Smith, *Lady Butler, Battle Artist, 1846–1933*, Sutton, 1987.

Kate Greenaway (1846–1901)

Born in Hoxton, London, to a working-class family, Catherine Greenaway is, of course, famous for her children's book illustrations.

Greenaway's parents home-schooled her; later she attended various dame schools; she was a voracious reader of chapbook editions of fairy tales as well as illustrated editions of Shakespeare, writing later that children 'often don't care a bit about the books people think they will and I think they often like grown-up books – at least I did.'

She trained in graphic design and art between 1858 and 1871 at Finsbury School of Art (night school classes from age 12 open only to women), There were four stages of courses, which she completed in 1864 before moving on to the Royal Female School of Art – here she was allowed to draw human figures, at first from plaster casts and then from models garbed in historical costumes – however, she was frustrated that there was no access to nude models in the women's classes so she enrolled in night school at Heatherley School of Fine Art where she met Edward Burne-Jones, Edward Poynter, and Walter Crane.

Unfortunately, women were segregated from men in the life class here too but, determined not to accept this discrimination, the South Kensington School of Art followed and then the Slade School of Fine Art came next when it opened in 1871 because they promised equal educational opportunities for men and women.

All the while, Greenaway was winning commissions for children's book illustrations. The first came in 1867 but 1869 brought the most significant when Frederick Warne & Co. purchased six illustrations for an edition of *Diamonds and Toads*. In 1868, she had her first exhibition at the famous Dudley Gallery in Piccadilly.

Her training complete, Greenaway started off designing for the burgeoning greetings card market, producing Christmas and Valentine's cards, the latter increasingly popular in this new age of romance. Freelancing for market leaders Marcus Ward & Co., early Valentines sold 25,000 copies in weeks, tapping in as they did into the booming floriography market. In 1879, when wood-block engraver and printer Edmund Evans printed *Under the Window*, an overnight bestseller, her reputation was assured.

Henrietta Eliza Vaughan Stannard (1856–1911)

Henrietta Stannard went by four different names: Henrietta Eliza Vaughan Stannard, her maiden name Palmer, the pen name Violet Whyte, and John Strange Winter.

In 1874, she assumed the *nom de plume* 'Violet Whyte', writing for the *Family Herald.* In 1881, she published *Cavalry Life*, a collection of regimental sketches, in 1883, *Regimental Legends*, and in 1885, *Bootles Baby*. All were written under the pseudonym of John Strange Winter, a character in one of her stories; her (male) publisher refused to bring out the books under a feminine pseudonym while the public assumed the author to be a cavalry officer. All in all, she published about 100 novels and ten other books as well as articles for the *Cornhill* and *Temple Bar*.

She married ex-soldier Arthur Stannard at Fulford, York in 1884. In 1885, *Bootle's Baby: A Story of the Scarlet Lancers* was published in the *Graphic*, selling two million copies in ten years. Between 1891 and 1895, she published her own weekly magazine and wrote *Confessions of a Publisher* in 1892, mocking patronising male publishers of the day. John Ruskin described John Strange Winter as 'the author to whom we owe the most finished and faithful rendering ever yet given of the character of the British soldier'. She was first president of the Writers' Club (1892), was a Fellow of the Royal Society of Literature and was president of the Society of Women Journalists from 1901 to 1903.

For a list of works, see www.victorianresearch.org/atcl/show_author.php?aid=758

Margaret MacDonald Mackintosh (1864–1933)

Painter and designer, she was a prominent member of the artist-designers later called the 'Glasgow Girls', working much of the time in collaboration with her sister, Frances MacDonald (1874–1921), James Herbert McNair (1868–1955) and Charles Rennie Mackintosh, her husband from 1900.

By 1890, the family was in Glasgow and Margaret and her sister enrolled as day students at the Glasgow School of Art for courses in design in which they worked with a variety of media, including metalwork, embroidery, and textiles. She also joined the Scottish Society of Watercolour Painters in 1898.

Mackintosh's *Seven Princesses* is her masterpiece.

In 2008, Mackintosh's 1902 work *The White Rose and the Red Rose*, a painted gesso over hessian with glass bead artwork, was auctioned for the equivalent of £2,866,260 in 2023.

It comes as no surprise to know that Margaret Mackintosh's talent has unfortunately been grossly overshadowed by her (admittedly) more famous husband.

Anna Alma-Tadema (1867–1943)

Anna Alma-Tadema was an artist and a suffragette; her *Miss Tessa Gosse* and fifteen or so other works, such as *The Misty Valley* and *The Gold Room*, were shown at the Royal Academy of Arts.

Alma-Tadema was devoted to women's suffrage and signed Helen Blackburn's *Some Supporters of the Women's Suffrage Movement* in 1897.

Following on from the sexist overshadowing we acknowledged with Margaret MacDonald, in 2022 Susie Beckam opened her article published in the *British Art Journal* with the following stinging salvo:

> Anna Alma-Tadema has been almost entirely excluded and ignored by art history. Where her work and life has been accorded minor attention by scholars, this consideration has frequently taken place within a wider and substantially more generous analysis of her father, Sir Lawrence Alma-Tadema (1836–1912). This has implied a narrative that she is only of interest to art history in the context of what she can reveal about a notable male family when, in fact, [she] was a highly regarded and widely exhibited artist in her own right who produced excellent work.

Epilogue

Anna Alma-Tadema's experience provides a fitting end to this parade of trailblazing women who were busy changing the world during the Industrial Revolution. She had to put up with being crowded out of history due to the adulation showered on her father. Margaret MacDonald was subjected to the same discrimination because of her husband's fame. And then there was the institutional sexism where revered institutions even changed their ancient constitutions to block out women and maintain their male bastion status; the sexism turned ugly at times with verbal abuse, even riots. Similar was the conviction that women simply couldn't and shouldn't be doing this, that or the other, it being man's work.

Misappropriation of intellectual property and plagiarism were never far away when women made important strides forward in various fields, nor was racism when enterprising women of colour emerged into a white, class-driven male world. Generally speaking, second-class citizenship was the best our trailblazing women could hope for, given the refusal they faced for many years relating to the right to vote, to compete on an even field in all the professions and to have equal opportunities to education and in the workplace.

But the good news is that these 200 or so women just rolled up their sleeves and got on with their trailblazing patiently and with determination; they opened doors hitherto slammed shut; they shattered glass ceilings and proceeded to leave their indelible stamp on our Industrial Revolution and its heritage, adding to and complementing the work done by their male counterparts and sometime adversaries. Reading the impressive and inspirational stories of these resolute and persistent women, we can't help believe that without them the huge change in our industrial base would have been somewhat less revolutionary than it turned out to be.

Bibliography

Aaron, Jane, *Nineteenth-Century Women's Writing in Wales: Nation, Gender and Identity*, Cardiff, 2010

Ablow, Rachel, '"One Flesh," One Person, and the 1870 Married Women's Property Act', *BRANCH: Britain, Representation and Nineteenth-Century History*, ed. Dino Franco Felluga. Extension of *Romanticism and Victorianism on the Net*

Alexander, Ziggi, 'Let it Lie Upon the Table: The Status of Black Women's Biography in the UK', *Gender & History*, 2(1), Spring 1990, pp. 22–33

Alexander, Ziggi, *Mary Seacole: Jamaican National Heroine and Doctress in the Crimean War*, Brent Library Service, 1982

Alexander, Ziggi, *Wonderful Adventures of Mrs Seacole in Many Lands*, Falling Wall Press, 1984

Allen, Robert C., *The British Industrial Revolution in Global Perspective*, Cambridge, 2009

Anderson, Christopher Gangadin, *London Vagabond: The Life of Henry Mayhew*, Amazon KDP, 2018

Anholt, Laurence, *Stone Girl Bone Girl: The Story of Mary Anning*, Frances Lincoln Publishers, 2006

Anionwu, Elizabeth N., *A Short History of Mary Seacole: A Resource for Nurses and Students*, Royal College of Nursing, 2005

Appleyard, Rollo, *The History of the Institution of Electrical Engineers*, London, Institution of Electrical Engineers, 1939

Armstrong, Benjamin, *Britain 1783–1885*, Hodder Education, 2020

Armstrong, Eva, 'Jane Marcet and her "Conversations on Chemistry"', *Journal of Chemical Education*, 15(2), 1938, pp. 53–57

Atkins, Jeannine, *Mary Anning and the Sea Dragon*, Farrar Straus Giroux, 1999

Atkinson, Blanche, *Ruskin's Social Experiment at Barmouth*, Sothis Press, 2023

Atkinson, Diane, *Rise Up, Women!: The Remarkable Lives of the Suffragettes*, London, 2018

Aveling, James Hobson, *English Midwives: Their History and Prospects*, J. & A. Churchill, 1872

Awcock, Hannah (29 December 2016), 'Turbulent Londoners: Barbara Leigh Smith Bodichon, 1827–1891', *Turbulent London*

Ayres, Brenda A. (ed.), *Frances Trollope and the Novel of Social Change*, Westport, CT, 2002

Ayrton, Hertha, *The Electric Arc*, London, 1902

Bacik, Ivana et al., *Gender Injustice: Feminising the Legal Professions?* (PDF). Trinity College Dublin Law School

Baker, Rachel, *The First Woman Doctor: The Story of Elizabeth Blackwell, M.D.*, New York, 1944

Balducci, Temma, 'Negotiating Identity: Mary Ellen Best and The Status of Female Victorian Artists', *Nineteenth-Century Art Worldwide*, 2015

Baum, Joan, *The Calculating Passion of Ada Byron*, Archon, 1986

Beckham, Susie, 'Recovering Anna Alma-Tadema (1867–1943)', *British Art Journal*, XXII(3), 2021–22, pp. 32–43

Bennett, Betty T., *Mary Wollstonecraft Shelley: An Introduction*, Baltimore, MD, 1998

Berg, Maxine, 'Women's Property and the Industrial Revolution', *The Journal of Interdisciplinary History*, 24(2), 1993, pp. 233–250

Bindslev, Anne M., *Mrs. Humphry Ward: A Study in Late-Victorian Feminine Consciousness and Creative Expression*, Stockholm, Almqvist & Wiksell International, 1985

Blackwell, Elizabeth and Marta McDowell (ed.). *A Curious Herbal: Elizabeth Blackwell's Pioneering Masterpiece of Botanical Art*, New York, 2023

Blackwell, Elizabeth, *The Influence of Women in the Profession of Medicine*, Baltimore, 1890

Blake, Catriona, *The Charge of the Parasols: Women's Entry to the Medical Profession*, The Women's Press Ltd., 1990

Borst, Charlotte G., *Catching Babies: Professionalization of Childbirth, 1870–1920*, Cambridge, MA, 1995

Bostridge, Mark, *Florence Nightingale: The Woman and Her Legend*, Viking, 2008

Boyd, Nancy, *Three Victorian Women Who Changed Their World: Josephine Butler, Octavia Hill, Florence Nightingale*, Oxford, 1982

Briggs, Julia, *A Woman of Passion: The Life of E. Nesbit, 1858–1924*. New Amsterdam Books, 1987

Brinks, Ellen, *Anglophone Indian Women Writers, 1870–1920*, Routledge, 2016

Broadbent, Lizzie, 'Laura Annie Willson (1877–1942)'. *Women Who Meant Business*, 2021

Brock, C., *British Women Surgeons and their Patients, 1860–1918*, Cambridge, 2017

Brockbank, Jonathan, 'York: A Virtual Cultural/Historical Tour: York in the late 18th early 19th Century: Female Creativity, Leisure and Class', University of York

Brown, Susan et al., 'Ellen Johnston', *Orlando*. orlando.cambridge.org, 2012. Web. 23 January 2013

Browne, Claire, *Silk Designs of the Eighteenth Century: From the Victoria and Albert Museum*, London, 1996

Bruton, Elizabeth, 'The life and material culture of Hertha Ayrton', *Science Museum Group Journal*, 10(10), 2018

Brück, Mary T, 'Mary Somerville, mathematician and astronomer of underused talents', *Journal of the British Astronomical* Association, 106(4), pp. 201–206

Buckland, Gail, *Fox Talbot and the Invention of Photography*, D. R. Godine, 1980

Burek, C. V. and Higgs, B. M., *Celebrating 100 Years of Female Fellowship of the Geological Society: Discovering Forgotten Histories*, Geological Society of London, 2021

Burek, Cynthia V., 'The first lady geologist, or collector par excellence?'. *Geology Today*, 17(5), 2001, pp. 192–194

Burkhauser, Jude (ed.), *Glasgow Girls: Women in Art and Design 1880–1920*, Edinburgh, 1990

Burnette, Joyce, 'The Wages and Employment of Female Day-Labourers in English Agriculture 1740–1850', *Economic History Review*, 57(4), 2004, pp. 664–690

Burnette, Joyce, 'Women Workers in the British Industrial Revolution', EH.Net, 2008

Burton, Antoinette, *At the Heart of the Empire: Indians and the Colonial Encounter in Late-Victorian Britain*, University of California Press, 1998, p. 115

Butler, Josephine Elizabeth Grey, *Woman's Work and Woman's Culture: A Series of Essays*, London, 1869

Cadbury, Deborah, *The Dinosaur Hunters: A True Story of Scientific Rivalry and the Discovery of the Prehistoric World*, Fourth Estate, 2000

Candlin, Kit (2016a), 'Thomas, Dorothy "Dolly" (c. 1763–1847), prominent free colored businesswoman and slave owner'. In: Knight, Franklin W. and Gates, Henry Louis Jr. (eds.), *Dictionary of Caribbean and Afro-Latin American Biography*, Oxford

Candlin, Kit (6 October 2016b), 'Thomas [formerly Kirwan], Dorothy', *Oxford Dictionary of National Biography* (online ed.), Oxford

Candlin, Kit and Pybus, Cassandra, 'The Queen of Demerara: Mrs. Dorothy Thomas'. In: *Enterprising Women: Gender, Race, and Power in the Revolutionary Atlantic*, Athens, Georgia, pp. 103–125, 2016

Carlson, J. A., *England's First Family of Writers: Mary Wollstonecraft, William Godwin, Mary Shelley*, Baltimore, MD, 2007

Carpenter, Humphrey, 'Kate Greenaway'. In: Carpenter and Prichard (eds.), *The Oxford Companion to Children's Literature*, New York, 1984

Cashman, Hilary, 'Singular Iniquities: Josephine Butler and Marietta Higgs', *New Blackfriars*, 71(834), 1990, pp. 26–32

Chapman, Alison, https://scottishwomenpoets.wordpress.com/poets/nineteenth-century-poets/ellen-johnston-the-factory-girl

Chapman, Allan, *Mary Somerville and the World of Science*, Bristol, 2004

Cherry, Deborah, *Beyond the Frame: Feminism and Visual Culture, Britain 1850–1900*, Routledge, 2012

Chiba, Akira, *Queen Victoria and the African Princess*, Museion Books

Chrystal, Paul, *A History of Sweets*, Pen & Sword, Barnsley, 2021

Chrystal, Paul, *A History of the World in 100 Pandemics, Plagues and Epidemics*, Pen & Sword, Barnsley, 2023

Chrystal, Paul, *Confectionery in Yorkshire*, Stroud, 2012

Chrystal, Paul, *Factory Girls and Climbing Boys: Women and Children at Work from the Industrial Revolution to 1900*, Pen & Sword, Barnsley, 2022

Chrystal, Paul, *History of Chocolate in York*, Pen & Sword, Barnsley, 2012

Collinson, S. R., 'Mary Ann Dacomb Scharlieb: A Medical Life from Madras to Harley Street', *Journal of Medical Biography*, 7(1), 1999, pp. 25–31

Combs, Mary Beth, '"A Measure of Legal Independence": The 1870 Married Women's Property Act and the Portfolio Allocations of British Wives', *The Journal of Economic History*, 65(4), pp. 1028–1057

Commire, Anne, *Women in World History: A Biographical Encyclopedia*, 2000, vii

Comrie, John D., Wellcome Historical Medical Museum, *History of Scottish medicine to 1860 [electronic resource]*. Wellcome Library, London, 1927

Cornforth, John (9 June 1983), 'An anticipation of Victoriana. The drawings of Mary Ellen Best', *Country Life*, pp. 1549–1552

Cotterell, T., *The Systematic Mineral Collection of Lady Henrietta Antonia Clive, Countess of Powis: A Rediscovered Early 19*th *Century Treasure*, 2014

Cowman, Krista, *The Routledge Companion to British Women's Suffrage*, London, 2024

Crawford, Elizabeth, *Enterprising Women: The Garretts and their Circle*, Francis Boutle Publishers, 2002

Crawford, Elizabeth, *The Women's Suffrage Movement: A Reference Guide 1866–1928*, London, Psychology Press, 1999

Creese, Mary R. S., *Ladies in the Laboratory? American and British Women in Science, 1800–1900: A Survey of Their Contributions to Research*, Scarecrow Press, 2000

Crellin, John K., 'Mrs. Marcet's "Conversations on Chemistry"', *Journal of Chemical Education*, 56(7), 1979, pp. 459–460

Crompton, Frank, *Workhouse Children: Infant and Child Paupers Under the Worcestershire Poor Law, 1780–1871*, Stroud, 1997

Cross, Wilbur L., 'Philosophical Realism: Mrs. Humphry Ward and Thomas Hardy'. In: *The Development of the English Novel*, New York, 1899

Danger, Sara R., 'Producing the Romance of Mass Childhood: Kate Greenaway's "Under the Window" and the Education Acts', *Nineteenth-Century Contexts*, 31(4), Dec. 2009, pp. 311–333

Davidoff, Leonore, 'Class and Gender in Victorian England: The Diaries of Arthur J. Munby and Hannah Cullwick', *Feminist Studies*, 5(1), 1979, pp. 87–141

Davidoff, Leonore, and Catherine Hall, *Family Fortunes: Men and Women of the English Middle Class, 1780–1850*, Chicago, University of Chicago Press, 1987

Davidson, Caroline, *The World of Mary Ellen Best*, London, 1985

De Montfort, Patricia, *Louise Jopling: A Biographical and Cultural Study of the Modern Woman Artist in Victorian Britain*, 2017

Devereux, Jo, *The Making of Women Artists in Victorian England: The Education and Careers of Six Professionals*, Jefferson, NC, 2016

Diary of Mrs Isabel Jane Thorne, campaigner for medical education for women, at the Wayback Machine (1887)

Dickens, Charles, *Mary Anning, the Fossil Finder*, *All Year Round*, 13, 1865

Dronsfield, Alan, 'Percivall Pott, chimney sweeps and cancer', *Education in Chemistry*, 43(2), pp. 40–42, 48

du Preez, Hercules Michael, 'Dr James Barry: The early years revealed', *SAMJ: South African Medical Journal*, 98(1), 2008, pp. 52–58

du Preez, Hercules Michael and Dronfield, Jeremy, *Dr James Barry: A Woman Ahead of Her Time*, London, Oneworld Publication, 2016

Dunn, Daisy, *Not Far From Brideshead: Oxford Between the Wars*, London, 2024

Dyhouse, Carol, 'Driving ambitions: women in pursuit of a medical education, 1890–1939', *Women's History Review*, 7(3), 1998, pp. 321–343

Edmunds, Percy James (18 March 1911), 'The Origin of the London School of Medicine for Women', *British Medical Journal, Correspondence*, 1(2620), p. 659

Ehrenreich, Barbara, *Witches, Midwives, and Nurses: A History of Women Healers*, The Feminist Press, 1973

English Heritage, 'Sarah Forbes Bonetta, Queen Victoria's African Protégée'

Erickson, Amy Louise, *Women and Property in Early Modern England*, London, 1993

Erickson, Amy Louise, 'Married women's occupations in eighteenth-century London', *Continuity & Change*, 23, 2008, 267–307

Essinger, James, *Ada's Algorithm: How Lord Byron's Daughter Ada Lovelace Launched the Digital Age*, Melville House Publishing, 2004

Evans-Pugh, Christine, 'Engineering's most ingenious women' (16 May 2011), www.eandt.theiet.org

Favret, Mary, *Romantic Correspondence: Women, Politics and the Fiction of Letters*, Cambridge, 1993

Feldman, Paula, R. (ed.), *British Women Poets of the Romantic Era: An Anthology*, Baltimore, MD, 1997, p. 712

Fillimore, Francis, 'Britain's Battle Painter: Lady Butler and Her Art', *New England Home Magazine*, XII (13), September 1900, pp. 579–587 (also published in *Windsor Magazine*, XI, December 1899–May 1900, pp. 643–652)

Finlay, R. B. and Buckmaster, Stanley (2 May 1908), 'The Admission of Women to the Royal College of Surgeons of England', *The British Medical Journal*, 1(2470), p. 1079

Fisch, Audrey A. (ed.), *The Other Mary Shelley: Beyond 'Frankenstein'*, New York

Ford, Lauren Moya (2 August 2023), 'The Exquisite Illustrations of a Pioneering Woman Herbalist', *Hyperallergic*

Ford, Steve (13 October 2015), 'Life and work of nurse Edith Cavell remembered', *Nursing Times*

Foster, John Bellamy, and Brett Clark, 'Women, nature, and capital in the industrial revolution', *Monthly Review*, 69(8), 2018, pp. 1–24

Frame, Paul, 'Clive, Henrietta Antonia (1758–1830), traveller and scientific collector', *Dictionary of Welsh Biography*, National Library of Wales, 2018

Fraser, Antonia, *The Case of the Married Woman: Caroline Norton: A 19*th *Century Heroine Who Wanted Justice for Women*, London, 2021

Fuegi, J. and Francis, J., 'Lovelace & Babbage and the creation of the 1843 "notes"' (PDF), *Annals of the History of Computing*, 25(4), 2003, pp. 16–26

Gates, Brian, 'Faith schools and colleges of education since 1800'. In Gardner, Roy, Lawton, Denis and Cairns, Jo (eds.), *Faith Schools: Consensus or Conflict?* Abingdon, 2005

Gaylor, Annie Laurie (ed.), *Women Without Superstition: 'No Gods – No Masters': The Collected Writings of Women Freethinkers of the Nineteenth and Twentieth Centuries*, Madison, Wisconsin, 1997

Girton College Register, 1869–1946

Godwin, William, *Memoirs of the Author of A Vindication of the Rights of Woman*, eds. Pamela Clemit and Gina Luria Walker, Peterborough, 2001

Goodhue, Thomas W., *Curious Bones: Mary Anning and the Birth of Paleontology* (Great Scientists), Morgan Reynolds, 2002

Goodhue, Thomas W., *Fossil Hunter: The Life and Times of Mary Anning (1799–1847)*, Academica Press, 2004

Gooptu, Suparna, *Cornelia Sorabji: India's Pioneer Woman Lawyer*, Oxford, 2006

Gordon, Charlotte, *Romantic Outlaws: The Extraordinary Lives of Mary Wollstonecraft and Mary Shelley*, Random House, 2016

Goulding, Christopher, 'The Real Doctor Frankenstein?' *Journal of the Royal Society of Medicine*, May 2002

Green, Stephanie, 'The Serious Mrs Stopes: Gender, Writing and Scholarship in Late-Victorian Britain', *Nineteenth-Century Gender Studies*, 5(3), 2009

Greenaway, Kate, *The Language of Flowers*, London, 1884

Griffin, Ben, 'Class, Gender, and Liberalism in Parliament, 1868–1882: The Case of the Married Women's Property Acts', *The Historical Journal*, 46(1), 2003, 59–87

Grinstein, Louise S. & Campbell, Paul J., *Women of Mathematics: A Bio-Bibliographic Sourcebook*, London, 1987

Guthrie, Douglas, *Extramural Medical Education in Edinburgh*, Edinburgh, 1965

Halsall, Paul (ed.), 'Women Miners in English Coal Pits 1842', *Internet Modern History Sourcebook*, 2016

Hamilton, Susan, 'Making History with Frances Power Cobbe', *Victorian Studies*, 43, 2001

Harry, Owen G., 'The Hon. Mrs Ward (1827–1869) Artist, Naturalist, Astronomer and Ireland's First Lady of the Microscope', *The Irish Naturalists' Journal*, 21(5), 1984, pp. 193–200

Hart, Alice, *Cottage Industries and What They Can Do for Ireland*, 1885

Hart, Monica Smith, 'The Factory Exile: Ellen Johnston's Autobiography, Poems and Songs', *Victorian Poetry*, 53(1), 2015, pp. 77–99

Hartley, Jenny, *Charles Dickens and the House of Fallen Women*, Methuen, 2008, p. 29

Hassan, Narin, *Diagnosing Empire: Women, Medical Knowledge, and Colonial Mobility*, Ashgate Publishing Ltd., 2011

Hayden, Ruth, Mrs Delany: Her Life and Her Flowers, London, 1980

Hayek, F. A., 'John Stuart Mill and Harriet Taylor: Their Friendship and Subsequent Marriage'. In: *Hayek on Mill: The Mill-Taylor Friendship and Related Writings*, Chicago, 2015

Heald, Henrietta, 'A forgotten feminist pioneer: the story of Rachel Parsons', *Electrifying Women*, 2020

Heald, Henrietta, 'Magnificent Women and their Revolutionary Machines', Unbound, 2021

Healey, Edna, *Lady Unknown: The Life of Angela Burdett-Coutts*, London, 1978

Heilmann, Ann, *Neo-/Victorian Biographilia and James Miranda Barry: A Study in Transgender and Transgenre*, London, 2018, pp. 65, 281–282

Helland, Janice, *The Studios of Frances and Margaret Macdonald*, Manchester, 1995, pp. 13–18

Helsinger, Elizabeth K., 'The Woman Question', *Social Issues, 1837–1883*, London, 1983

Henley, Dorothy, *Rosalind Howard, Countess of Carlisle*, Hogarth Press, 1958

'Henrietta Vansittart – Female engineer of the Industrial revolution', *literarylandscapes*, 9 March 2017

'Hertha Ayrton: pioneering inventor and suffragette', *Physics World*, 6 September 2022

Higginbotham, Peter, 'An Introduction to the Workhouse', workhouses.org.uk, 2013

Hill, Bridget, *Women, Work and Sexual Politics in Eighteenth-Century England*, London, 1988

Hill, Michael R., *Harriet Martineau: Theoretical and Methodological Perspectives*, Routledge, 2002

Hirsch, Pam, *Barbara Leigh Smith Bodichon: Feminist, Artist and Rebel*, London, 1998

Hoehling, A. A., 'The Story of Edith Cavell', *The American Journal of Nursing*, 57(10), Lippincott Williams & Wilkins, 1957, pp. 1320–1322

Holland, Janice, 'Working bodies, Celtic textiles and the Donegal Industrial Fund, 1883–90', *Textile: The Journal of Cloth and Culture*, ii(2) (July 2004), 134–155

Holland, Janice, *British and Irish Home Arts and Industries, 1880–1914: Marketing Craft, Making Fashion*, Irish Academic Press, 2007

Holmes, Rachel, *Scanty Particulars: The Scandalous Life and Astonishing Secret of James Barry, Queen Victoria's Most Eminent Military Doctor*, London, 2003

Holmes, Rachel, *The Secret Life of Dr James Barry: Victorian England's Most Eminent Surgeon*, London, 2020

Honeyman, Katrina, *Women, Gender and Industrialisation in England, 1700–1870*, New York, St Martin's Press, 2000

Horn, Jeff, *The Industrial Revolution*, Greenwood, 2007

Horrell, Sara, and Jane Humphries, 'Women's Labour Force Participation and the Transition to the Male-Breadwinner Family, 1790–1865', *Economic History Review*, 48 (1995), pp. 89–117

Horrocks, Sally (29 April 2016), 'Trailblazers of diversity', *The Engineer*

Howsam, Leslie, *Eliza Orme's Ambitions: Politics and the Law in Victorian London*, Cambridge, 2024

https://mathshistory.st-andrews.ac.uk/Biographies/Ayrton

https://womenwhomeantbusiness.com/2021/06/05/louise-jopling-1843-1933

Hudson, Derek, *Munby: Man of Two Worlds. The Life and Diaries of Arthur J. Munby, 1828–1910*, London, 1974

Hume, Leslie Parker, 'The National Union of Women's Suffrage Societies, 1897–1914', *Modern British History*, 3, New York, 1982

Humphries, Jane, *Childhood and Child Labour in the British Industrial Revolution*, Cambridge, 2011

Hunt, Karen, *Equivocal Feminists*, Cambridge, 1996

Innes, C. L., *A History of Black and Asian Writers in Britain*, Cambridge, 2008

Jacobs, Jo-Ellen, *The Voice of Harriet Taylor Mill*, Indiana University Press, 2002

Jaffe, Deborah, *Ingenious Women*, Stroud, 2003

Jayawardena, Kumari, *The White Woman's Other Burden: Western Women and South Asia During British Rule*, London, 1995

John, Angela, *By the Sweat of Their Brow: Women Workers at Victorian Coal Mines*, London, 1980

Johnson, James, 'Women Inventors and Discoverers', *Cassier's Magazine*, 1909, pp. 548–553

Johnston, Ellen, *Autobiography, Poems and Songs of Ellen Johnston, the 'Factory Girl'*, W. Love, 1867

Johnston, Johanna, *The Life, Manners, and Travels of Fanny Trollope*, 1979

Jones, Vivien, 'The Death of Mary Wollstonecraft', *BritishJournal for Eighteenth-Century Studies*, 20(2), 1997, pp. 187–205

Jopling, Louise, 'Twenty Years of My Life: 1867–1887', 1925

Jordan, Alison, *Margaret Byers: Pioneer of Women's Education and Founder of Victoria College, Belfast*, Belfast, 1991

Jordan, Jane, *Josephine Butler*, Hambledon Continuum, 2001

Jordan, Thomas E., 'The Keys of Paradise: Godfrey's Cordial and Children in Victorian Britain', *Journal of the Royal Society of Health*, 107(1), 1987, pp. 19–22

Kaufman, Matthew H., *Medical Teaching in Edinburgh During the 18*th *and 19*th *Centuries*, Royal College of Surgeons of Edinburgh, 2003

Kelly, Alison, 'Coade Stone in Georgian Gardens', *Garden History*, 16(2), 1988, pp. 109–133

Kelly, Laura, 'Elizabeth Garrett Anderson: early pioneer of women in medicine' (PDF), *The Lancet*, 390(10113), 2017, pp. 2620–2621

Kelly, Matthew, *The Women Who Saved the English Countryside*, Yale University Press, 2022

Kerhervé, Alain (ed.), *Mary Delany (1700–1788) and the Court of King George III*, vol. 4 of Michael Kassler (ed.), *Memoirs of the Court of George III*, London, 2015

Kim, Eugene and Toole, Betty Alexandra, 'Ada and the First Computer', *Scientific American*, 280(5), 1999, pp. 66–71

Kissel, Susan S., *In Common Cause: The 'Conservative' Frances Trollope and the 'Radical' Frances Wright*, Bowling Green, OH, 1993

Klaus, H. Gustav, *Factory Girl: Ellen Johnston and Working-class Poetry in Victorian Scotland*, Peter Lang, 1998

Koerner, Emily Rees, 'Henrietta Vansittart', *Science Museum Group Journal*, 15(15), 2021

Kubba, A. K. and Young, M., 'The Life, Work and Gender of Dr James Barry MD', *Proceedings of the Royal College of Physicians of Edinburgh*, 31(4), 2001, pp. 352–356

Kujawinski, Bethany, 'Princess Omoba Aina (Sarah Forbes Bonetta) of Nigeria', editions.covecollective.org, 2020

Laird, Mark and Weisberg-Roberts, Alicia (eds.), *Mrs. Delany and Her Circle*, New Haven, 2009

Lalumia, Matthew Paul, 'Lady Elizabeth Thompson Butler in the 1870s', *Woman's Art Journal*, 4(1), Spring–Summer 1983

Larmour, Paul, 'The Donegal industrial fund and its cottage industries', *GPA Arts Review Yearbook (1990–91)*, pp. 128–134

Laufer, Geraldine Adamich, *Tussie-Mussies: The Victorian Art of Expressing Yourself in the Language of Flowers*, Workman Publishing, 1993, pp. 4–25, 40–53

Layne, Margaret E., *Women in Engineering: Pioneers and Trailblazers*, Reston, VA, 2009

Lear, Linda, *Beatrix Potter: A Life in Nature*, London, 2007

Lear, Linda, *Beatrix Potter: The Extraordinary Life of a Victorian Genius*, London, 2008

Ledger, Sally, *The New Woman: Fiction and Feminism at the Fin de Siècle*, Manchester University Press, 1997

Leneman, Leah, 'The awakened instinct: vegetarianism and the women's suffrage movement in Britain', *Women's History Review*, 6(2), 1997, pp. 271–287

Lewis, Judith S., 'Princess of Parallelograms and her daughter: Math and gender in the nineteenth century English aristocracy', *Women's Studies International Forum*, 18(4), 1995, pp. 387–394

Limoncelli, Stephanie, *The Politics of Trafficking: The First International Movement to Combat the Sexual Exploitation of Women*, Stanford, CA, 2010

Logan, D. A. (ed.), *The Collected Letters of Harriet Martineau*, London, 2007

Lomax, Elizabeth, 'The Uses and Abuses of Opiates in Nineteenth-Century England', *Bulletin of the History of Medicine*, 47(2), 1973, pp. 167–176

Lovejoy, Esther Pohl, *Women Doctors of the World*, London, 1957

Lucey, John, 'Ladies of the Law: Pioneering Women in Irish Legal History', *History Ireland*, 29(1), 2021, pp. 46–49

Lyndon Shanley, Mary, *Feminism, Marriage, and the Law in Victorian England*, Princeton University Press, 1993

Madge, Bruce (30 July 2001), 'Elizabeth Blackwell – the forgotten herbalist?' *Health Information & Libraries Journal*, 18(3), pp. 144–152

Maggs, Sam, *Wonder Women: 25 Innovators, Inventors, and Trailblazers who Changed History*, Quirk Books, 2016

Manning, R., 'The Tyranny of Bodily Strength: Harriet Taylor Mill and John Stuart Mill on Domestic Violence'. In: *Analyzing Violence Against Women*, W. Teays (ed.), Cham, Switzerland, 2019

Manton, Jo, *Elizabeth Garrett Anderson: England's First Woman Physician*, London, 1965

Marcus, Sharon, *Between Women: Friendship, Desire, and Marriage in Victorian England*, Princeton, 2009

Married Women's Property Act 1882, legislation.gov.uk. UK Public General Acts

Mason, Joan, 'Hertha Ayrton and the Admission of Women to the Royal Society of London'. In: *Notes and Records of the Royal Society*, London, Royal Society Publishing, 1991

McDonald, Lynn, 'Florence Nightingale and Mary Seacole on Nursing and Health Care', *Journal of Advanced Nursing*, 69(11), 2013

McCabe, Helen, '"Political … civil and domestic slavery": Harriet Taylor Mill and Anna Doyle Wheeler on marriage, servitude, and socialism', *British Journal for the History of Philosophy*, published online 4 May 2020

McDonald, Lynn, 'Florence Nightingale and Mary Seacole: Which is the Forgotten Hero of Health Care and Why', *Scottish Medical Journal*, 59(1), 2014, pp. 66–69

McDonald, Lynn, 'Wonderful Adventures: How Did Mary Seacole Come To Be Viewed as a Pioneer of Modern Nursing?' *Times Literary Supplement* (6 December 2013), pp. 14–15

McDonald, Lynn, *Mary Seacole: The Making of the Myth*, Iguana, 2014

McGowan, Christopher, *The Dragon Seekers*, Persius Publishing, 2001

McIntyre, N., 'Britain's first medical marriage: Frances Morgan (1843–1927), George Hoggan (1837–1891) and the mysterious "Elsie"', *Journal of Medical Biography*, 12(2), 2004, 105–114

McVicker, Mary F., *Women Adventurers, 1750–1900: A Biographical Dictionary, with Excerpts from Selected Travel Writings*, 2013

Melander, Eric, 'Welfare Cuts and Crime: Evidence from the New Poor Law', *The Economic Journal*, 133(651), 2022, 1248–1264

Mellor, Anne K., *Mary Shelley: Her Life, her Fiction, Her Monsters*, London, 1990

Mesnard, Élise-Marie, *Miss E. Blackwell et les femmes Médecins*, Bordeaux, 1889

Mitchell, Sally, *Frances Power Cobbe: Victorian Feminist, Journalist, Reformer*, University of Virginia Press, 2004

Mitchell, Sally, 'Ephemeral Journalism and Its Uses: Lucie Cobbe Heaton Armstrong (1851–1907)', *Victorian Periodicals Review*, 42(1), 2009, pp. 81–92

Mitchell, W. R., *Beatrix Potter: Her Lakeland Years*, Great Northern Books Ltd, 2010

Molinari, Veronique, '"Schools of their own" – The Ladies' Medical College and London School of Medicine for Women'. In: Andreolle and Molinari (eds.), *Women and Science, 17*th *Century to Present: Pioneers, Activists and Protagonists*. Part II, Chapter 2. Cambridge Scholars Publishing, 2011, pp. 99–124.

Moore, Glenis, 'Hertha Ayrton – first lady of the IEE', *Electronics and Power*, 32(8), 1986, pp. 583

Morantz, Regina Markell, 'Feminism, Professionalism, and Germs: The Thought of Mary Putnam Jacobi and Elizabeth Blackwell', *American Quarterly*, 34(5), 1982, 459–478

Morantz-Sanchez, Regina, 'Feminist Theory and Historical Practice: Rereading Elizabeth Blackwell', *History and Theory*, 31(4), 1992

Mossman, Mary Jane, *The First Women Lawyers: A Comparative Study of Gender, Law and the Legal Professions*, Bloomsbury Publishing, 2006

Münch, Ole, 'Henry Mayhew and the Street Traders of Victorian London – A Cultural Exchange with Material Consequences', *The London Journal*, 2017

Myers, Mitzi, 'Godwin's *Memoirs* of Wollstonecraft: The Shaping of Self and Subject', *Studies in Romanticism*, 20, 1981, pp. 299–316

Myers, Walter Dean, *At Her Majesty's Request: An African Princess in Victorian England*, New York, 1999

Neat, Timothy, 'Tinker, Tailor, Soldier, Sailor: Margaret Macdonald and the Principle of Choice'. In: Burkhauser, Jude (ed.), *Glasgow Girls*, 1990

Neville-Sington, Pamela, *Fanny Trollope: The Life and Adventures of a Clever Woman*, Viking, 1997

Nicholas, Stephen, 'The Living Standards of Women during the Industrial Revolution, 1795–1820', *The Economic History Review*, 46(4), 1993, pp. 723–749

Nightingale, Florence, *Notes on Matters Affecting the Health, Efficiency, and Hospital Administration of the British Army*, London, 1858

Nightingale, Florence, *Workhouse Nursing*, London, 1867

Nimura, Janice P., *The Doctors Blackwell: How Two Pioneering Sisters Brought Medicine to Women and Women to Medicine*, W. W. Norton & Company, 2021

Nixon, Cheryl L., *The Orphan in Eighteenth-Century Law and Literature*, Ashgate Publishing, 2011

Norman, David B., 'Mary Anning and her times: the discovery of British palaeontology (1820–1850)', *Trends Ecol. Evol.*, 14(11), 1999

Norton-Smith, J., 'An Introduction to Mrs. Humphry Ward, Novelist', *Essays in Criticism*, 18(4), 1968, pp. 420–428

Obermayer, D. (6 June 2019), 'Who was "James Miranda Stuart Barry"?' *Notes on a Gentleman*

Ogilvie, Marilyn Bailey, *Women in Science: Antiquity through Nineteenth Century: A Biographical Dictionary*, Massachusetts, MIT, 1990

Oldfield, Sybil, *Women Humanitarians: A Biographical Dictionary of British Women Active Between 1900 and 1950: 'Doers of the Word'*, Continuum, 2001

Orton, Diana, *Made of Gold: A Biography of Angela Burdett-Coutts*, Hamish Hamilton, 1980

Parssinen, Terry M., *Secret Passions, Secret Remedies: Narcotic Drugs in British Society, 1820–1930*, Manchester, 1983

Phillips, Nicola, 'Eleanor Coade', May 2009 at the Wayback Machine, Addidi Inspiration Award for Female Entrepreneurs

Pickford, Susan, '"I have no pleasure in collecting for myself alone": Social Authorship, Networks of Knowledge, and Etheldred Benett's Catalogue of the Organic Remains of the County of Wiltshire (1831)' (PDF), *Journal of Literature and Science*, 8, 2015, pp. 69–85

Pinchbeck, Ivy, *Women Workers and the Industrial Revolution, 1750–1850*, London, Routledge, 1930

Pirie, Mary, *Familiar Teachings on Natural History: A Book for the Use of Schools and Families*, 1864, Lewis Smith

Pirie, Mary, *Flowers, Grasses, and Shrubs: A Popular Book on Botany*, 1860

Power, John Carroll, *The Rise and Progress of Sunday Schools: A Biography of Robert Raikes and William Fox*, Sheldon, 1863, p. 240

Pringle, Rosemary, *Sex and Medicine: Gender, Power and Authority in the Medical Profession*, Cambridge, 1998

Proffitt, Pamela, *Notable Women Scientists*, Detroit, 1999

Quirk, Maria, 'Portraiture and Patronage: Women, Reputation and the Business of Selling Art, 1880–1914', *Visual Culture in Britain*, 17(2), 2016, pp. 181–199

Rae, Isobel, *The Strange Story of Dr James Barry*, 1958

Ramdin, Ron, *Mary Seacole*, Haus Publishing, 2005

Rappaport, Helen, *Encyclopedia of Women Social Reformers*, Volume 1, ABC-CLIO, 2001

Rappaport, Helen, *In Search of Mary Seacole: The Making of a Cultural Icon*, Simon and Schuster, 2022

Rappaport, Helen, *No Place for Ladies: The Untold Story of Women in the Crimean War*, Arum, 2007

Rayner-Canham, Marelene and Rayner-Canham, Geoff (24 February 2022), 'Pioneers of the London School of Medicine for Women (1874–1947): Their Contributions and Interwoven Lives', *Springer Nature*

Rayner-Canham, Marelene and Rayner-Canham, Geoff, 'Chemistry Was Their Life: Pioneering British Women Chemists, 1880–1949', *World Scientific*, 2008

Rees, Emily, 'The Long Read: Discovering the Victorian Engineer Henrietta Vansittart, part 1', *Electrifying Women*, 2020

Richardson, Sarah, 'You Know Your Father's Heart: The Cobden Sisterhood and the Legacy of Richard Cobden'. In: Howe, Anthony and Morgan, Simon (eds.), *Nineteenth Century Liberalism: Richard Cobden Bicentenary Essays*, Aldershot, 2006

Riedesel, Paul, L. 'Who Was Harriet Martineau?', *Journal of the History of Sociology*, 3, 1981, pp. 63–80

Rippon, Matthew J., 'What Is the Geography of Geographical Indications? Place, Production Methods and Protected Food Names', *Area*, 46(2), 2014, 154–162

Roberts, Charles, *The Radical Countess*, Steel Bros, 1962

Roberts, Shirley, *Sophia Jex-Blake*, London, 1993

Roberts, Shirley, 'Sophia Jex-Blake', *Oxford Dictionary of National Biography*, 2004

Robinson, Jane, *Barbara Leigh Smith Bodichon, Trailblazer – The First Feminist to Change Our World*, London, 2024

Robinson, Jane, *Mary Seacole: The Charismatic Black Nurse Who Became a Heroine of the Crimea*, Constable, 2004

Rogers, Joseph, *Reminiscences of a Workhouse Medical Officer*, T. F. Unwin, 1889

Rogers, Sarah, 'The Nurses of the 1897 Maidstone Typhoid Epidemic: Social Class and Training. How representative were they of mid-nineteenth century nursing reforms?' Unpublished Master of Letters dissertation, Dundee, March 2016

Rose, Lionel, *The Erosion of Childhood: Childhood in Britain 1860–1918*, London, 2002

Rosen, Judith, 'Class and Poetic Communities: The Works of Ellen Johnston, "The Factory Girl"', *Victorian Poetry*, 39(2), 2001, 207–228

Rosenfeld, Louis, 'The Chemical Work of Alexander and Jane Marcet' (PDF), *Clinical Chemistry*, 47(4), 2001, pp. 784–792

Rossi, A., 'Sentiment and Intellect: The Story of John Stuart Mill and Harriet Taylor Mill'. In: *Essays on Sex Equality*, Chicago, University of Chicago Press, 1970

Rossotti, Hazel, 'The Woman That Inspired Faraday', *Chemistry World*, 2007, pp. 58–61

Rothstein, Natalie, *The Victoria and Albert Museum's Textile Collection: Woven Textile Design in Britain to 1750*, New York, 1994

Sage Journals website, 'Problems of marriage and sexual morality: the Lambeth Conference: Mary Scharlieb', published 21 July 2020

Saini, Adarsh, 'Family and Gender in the Industrial Revolution', n.d.

Sanders, Mike (ed.), *Women and Radicalism in the Nineteenth Century: Frances Wright*, Vol. II, New York, 2001

Schaaf, Larry, 'The First Photographically Printed and Illustrated Book', *The Papers of the Bibliographical Society of America*, 73(2), 1979, p. 212

Scott, Nicola, 'Jex-Blake, Sophia', *The New Biographical Dictionary of Scottish Women*, Edinburgh, 2018, pp. 218–219

Shanley, Mary Lyndon, 'Suffrage, Protective Labor Legislation, and Married Women's Property Laws in England', *Signs*, 12(1), 1986, pp. 62–77

Shanley, Mary Lyndon, *Feminism, Marriage, and Law in Victorian England, 1850–1895*, Princeton, 1989

Shelley, Chris et al., *Industrialisation and Social Change in Britain*, 2016

Shields, Nancy K., *Birds of Passage: Henrietta Clive's Travels in South India 1798–1801*, Eland, 2001

Shopland, Norena, *A History of Women in Men's Clothes: From Cross-dressing to Empowerment*, Barnsley, 2021

Somerville, J. M., 'Dr Sophia Jex-Blake and the Edinburgh School of Medicine for Women, 1886-1898', *J R Coll Physicians Edinb.*, 35(3), 2005, 261–267

Sorabji, Richard, *Opening Doors: The Untold Story of Cornelia Sorabji, Reformer, Lawyer and Champion of Women's Rights in India*, Penguin Books India, 2010, pp. 8–16

Sorabji, Cornelia, *India Calling: The Memories of Cornelia Sorabji*, Nisbet, 1934, pp. 7–8
Spence, Margaret E., 'Ruskin's Friendship with Mrs Fanny Talbot', *Bulletin of the John Rylands Library Manchester*, 1960
St Clair, William, *The Godwins and the Shelleys: The Biography of a Family*, New York, 1989
Stanford, Caroline, 'Revisiting the Origins of Coade Stone' (PDF), *The Georgian Group Journal*, 24, 2016, pp. 95–116
Stanley, A., *Mothers and Daughters of Invention: Notes for a Revised History of Technology*, Rutgers University Press, New Brunswick, NJ, 1995
Stein, Dorothy, *Ada: A Life and a Legacy*, Cambridge, MA, 1985
Stone, Alison, 'Frances Power Cobbe and the Philosophy of Anti-Vivisection' (PDF), *Journal of Animal Ethics*, 13(13), 2023, pp. 21–30
Strange, Kathleen H., *Climbing Boys: A Study of Sweeps' Apprentices 1772–1875* (PDF), London, 1982
Strange, Kathleen H., *The Climbing Boys*, London, 1985
Strathern, Paul, *Quacks, Rogues and Charlatans*, Royal College of Physicians, London, 2015
Strickland, Elisabetta, *The Ascent of Mary Somerville in 19*th *Century Society*, Springer, 2016
Strobel, Heidi, *The Art of Mary Linwood: Embroidery, Installation and Entrepreneurship in Britain 1787–1845*, London, 2024
Summers, Anne, '"The Constitution Violated": The Female Body and the Female Subject in the Campaigns of Josephine Butler', *History Workshop Journal*, 48(48), 1999, pp. 1–15
Sutherland, Gill, *Faith, Duty, and the Power of Mind: The Cloughs and Their Circle, 1820–1960*, Cambridge, 2006
Taithe, Bertrand, *The Essential Mayhew: Representing and Communicating the Poor*, Rivers Oram Press, 1996
Taylor, Ann, *The Autobiography and Other Memorials of Mrs Gilbert, Formerly Ann Taylor*, ed. Josiah Gilbert. London: Henry S. King & Co., 1874
Taylor, Ina, *The Art of Kate Greenaway: A Nostalgic Portrait of Childhood*, Gretna, LA, 1991
Taylor, Jane, *Memoirs, Correspondence and Poetical Remains of Jane Taylor*, Volume I of *The Writings of Jane Taylor, In Five Volumes*, Boston: Perkins & Marvin, 1832
Taylor, Judy, *Beatrix Potter: Artist, Storyteller and Countrywoman* (revised ed.), F. Warne & Co., 1996
Taylor, Judy, *That Naughty Rabbit: Beatrix Potter and Peter Rabbit*, Warne, 2002
Terry, R. C. (ed.), *Oxford Reader's Companion to Trollope*, Oxford, 1999
Thomson, Susan, *Manchester's Victorian Art Scene and Its Unrecognised Artists*, Manchester Art Press, 2007
Thomson, Susan, *The Life and Works of Annie Louisa Swynnerton*, Manchester Art Press, 2018
Thorne, Isabel, 'Employment for Educated Women', *Englishwoman's Review of Social and Industrial Questions* [1867], 37(271), 1906, pp. 257–258
Tilly, Louise A., 'Women, Women's History, and the Industrial Revolution', *Social Research*, 61(1), 1994, pp. 115–137
Todd, Janet, 'Mary Wollstonecraft and the Rights of Death', *Gender, Art and Death*, Cambridge, 1993
Todd, Margret, 'The Life of Sophia Jex-Blake', *Nature*, 101(2546), 1918, pp. 461–462. Available at the Internet Archive
Tomalin, Claire, *The Life and Death of Mary Wollstonecraft* (revised ed.), New York, 1992
Toole, Betty Alexandra, *Ada, The Enchantress of Numbers: A Selection from the Letters of Ada Lovelace, and her Description of the First Computer*, Strawberry Press, 1992

Toole, Betty Alexandra, *Ada, The Enchantress of Numbers: Prophet of the Computer Age*, Strawberry Press, 1998

Torrens, Hugh, 'Mary Anning (1799–1847) of Lyme; "The Greatest Fossilist the World Ever Knew"', *The British Journal for the History of Science*, 25(3), 1995, pp. 257–284

Towns, Elmer L., 'History of Sunday School', *Sunday School Encyclopedia*, 1993

Twining, Louisa, *Workhouses and Women's Work*, London, 1858

Ulrich, Laurel Thatcher, *A Midwife's Tale: The Life of Martha Ballard Based on Her Diary, 1785–1812*, 1991

Valenze, Deborah, *The First Industrial Woman*, Oxford, 1995

van Lieshout, C., et al., 'Female entrepreneurship: business, marriage and motherhood in England and Wales, 1851–1911', *Social History*, 44(4), 2019, pp. 440–468

Wade, Stephen, *The Justice Women: The Female Presence in the Criminal Justice System 1800–1970*, Pen and Sword, 2015

Waldron, H. A., 'A brief history of scrotal cancer', *British Journal of Industrial Medicine*, 40(4), 1983, pp. 390–401

Walkowitz, Judith R., *Prostitution and Victorian Society: Women, Class, and the State*, Cambridge, 1982a

Walkowitz, Judith R., 'Male Vice and Feminist Virtue: Feminism and the Politics of Prostitution in Nineteenth-Century Britain', *History Workshop*, 13(13), pp. 79–93, 1982b

Walsh, Mary Roth, *Doctors Wanted: No Women Need Apply: Sexual Barriers in the Medical Profession, 1835–1975*, 1977

Warren, James, 'Three Unsung Inventions: Tarmac, Time and Propellers', Open University

Watson, William Henry (22 August 1862), 'The first fifty years of the Sunday school', London, Sunday School Union

Webb, Robert K., *Harriet Martineau, a Radical Victorian*, London, 1960

Weiner, Gaby, 'Harriet Martineau: A reassessment (1802–1876)'. In: Dale Spender (ed.), *Feminist Theorists: Three Centuries of Key Women Thinkers*, Pantheon, 1983, pp. 60–74

Wilkes, Sue, 'A Working Woman's Place', *History Today*, 67(6), June 2017

Williamson, Lori, *Power and Protest: Frances Power Cobbe and Victorian Society*, Rivers Oram, 2004

Wills, Matthew (10 September 2018), 'The Bizarre Victorian Diaries of Cullwick and Munby', *JSTOR Daily*

Wills, Stella, 'The Anglo-Jewish Contribution to the Education Movement for Women in the Nineteenth Century', *Transactions Jewish Historical Society of England*, 17, Jewish Historical Society of England, 1951–52, pp. 269–281

Wilson, Dorothy Clarke, *Lone Woman: The Story of Elizabeth Blackwell, the First Woman Doctor*, Little Brown, Boston, 1970

Wohl, Anthony S., 'Opium and Infant Mortality', 1989, https://victorianweb.org/science/health/health4.html

Wojtczak, Helena, 'The National Union of Women's Suffrage Societies', *Victorian Web*, 2000

Wright, Mary, 'Elizabeth Blackwell of Bristol: The First Woman Doctor', Bristol Historical Association pamphlets, no. 87, 1995

You, Xuesheng, 'The missing half: female labour force participation in Victorian England and Wales', in *The Online Historical Atlas of Occupational Structure and Population Geography in England and Wales 1600–2011*, ed. L. Shaw-Taylor, 2017

You, Xuesheng, 'Women's labour force participation in nineteenth-century England and Wales: evidence from the 1881 census enumerators' books', *Economic History Review*, 73(1), 2020, 106–133

Index